KML & KMZ

SURVEYING MATHEMATICS MADE SIMPLE

An original Book by

Jim Crume P.L.S., M.S., CFedS

Co-Authors
Cindy Crume
Bridget Crume
Troy Ray R.L.S.
Mark Sandwick P.L.S.
Mark Lull

KINDLE - PRINTED EDITIONS

*****

PUBLISHED BY:

Jim Crume P.L.S., M.S., CFedS

KML & KMZ

Copyright 2015 © by Jim Crume P.L.S., M.S., CFedS

All Rights Reserved

First publication: October, 2015

Cover photo: Courtesy of NASA.gov

KML & KMZ

TERMS AND CONDITIONS

The content of the pages of this book is for your general information and use only. It is subject to change without notice.

Neither we nor any third parties provide any warranty or guarantee as to the accuracy, timeliness, performance, completeness or suitability of the information and materials found or offered in this book for any particular purpose. You acknowledge that such information and materials may contain inaccuracies or errors and we expressly exclude liability for any such inaccuracies or errors to the fullest extent permitted by law.

Your use of any information or materials in this book is entirely at your own risk, for which we shall not be liable. It shall be your own responsibility to ensure that any products, services or information available in this book meet your specific requirements.

This book is covered by the Kindle Direct Publishing and/or CreateSpace Terms and Conditions.

This book may not be further reproduced or circulated in any form, including email. Any reproduction or editing by any means mechanical or electronic without the explicit written permission of Jim Crume is expressly prohibited.

KML & KMZ

TABLE OF CONTENTS

INTRODUCTION ------------------------------------4
DEFINITIONS --------------------------------------6
GRID ADJUSTMENT FACTOR (GAF) -------------8
REPRESENTATIVE PARCEL ---------------------12
TRAVERSE PC ------------------------------------17
MICROSTATION ----------------------------------30
CIVIL 3D ---49
TRIMBLE BUSINESS CENTER ------------------66
ABOUT THE AUTHOR --------------------------78

KML & KMZ

INTRODUCTION

Straight forward Step-by-Step instructions.

This book is just one part in a series of digital and paperback books on Surveying Mathematics Made Simple. The subject matter in this book will utilize the methods and formulas that are covered in the books that precede it. If you have not read the preceding books, you are encouraged to review a copy before proceeding forward with this book.

For a list of books in this series, please visit:

http://www.cc4w.net/ebooks.html

Prerequisites for this book:

A basic knowledge of "Traverse PC©, MicroStation©, Civil 3D© or Trimble Business Center©", Google Earth© and Grid Adjustment Factor (a.k.a. combined factor) (see below) are required.

Traverse PC© is copyrighted by Traverse PC, Inc. Traverse PC screen shots reprinted courtesy of Traverse PC, Inc.

MicroStation© is copyrighted by Bentley Systems, Inc. MicroStation screen shots reprinted courtesy of Bentley Systems, Inc.

Civil 3D® is copyrighted by Autodesk, Inc. Civil 3D screen shots reprinted courtesy of Autodesk, Inc.

KML & KMZ

Trimble Business Center© (TBC) is copyrighted by Trimble Navigation Limited. TBC screen shots reprinted courtesy of Trimble Navigation Limited.

Google Earth© is copyrighted by Google, Inc. Google Earth screen shots reprinted courtesy of Google, Inc.

This book is not intended to be a complete tutorial on the use of the programs discussed in the pages of this book.

This book is only focusing on the basic KML & KMZ features associated with each of the programs shown herein.

It is noted that the programs shown herein may have advanced KML & KMZ features that are not demonstrated.

You are encouraged to experiment with those features and to consult the user manuals.

KML & KMZ

DEFINITIONS

Google Earth (GE): Google Earth comes in two versions. Google Earth (limited resolution images) and Google Earth Pro (High resolution images). As of the writing of this book, Google Earth Pro is FREE to download and install. We recommend Google Earth Pro for the high resolution images for printing and saving.

Grid Adjustment Factor (GAF): Also known as the Combined Factor. This factor is a combination of the Grid Factor and Ellipsoid Factor for converting from Grid Coordinates to Ground Coordinates and vice versa.

KML: Keyhole Markup Language (KML) is an XML notation for expressing geographic annotation and visualization. KML was developed for use with Google Earth. KML became the international standard of the Open Geospatial Consortium in 2008. (https://en.wikipedia.org/wiki/Keyhole_Markup_Language)

KMZ: Is a compressed (ZIP) file that contains a main KML file and as needed, supporting files. Google Earth will read KML or KMZ files directly.

 Throughout this book, tips will be given to help explain or give directions on the subject matter.

 This Icon is for CADD tips or directions that will be given throughout this book.

Reference to SimPro-GC Suite© and/or it's modules.

SimPro-GC is an acronym for **Sim**ple **Pro**gram **G**raphical **C**oordinate program. This is a custom Windows program designed by the author of this book that interfaces with MicroStation for performing coordinate geometry, project management as well as other functions.

This program is not needed for examples shown in this book.

GRID ADJUSTMENT FACTOR (GAF)

a.k.a. Combined Factor

A single project GAF is commonly used in the surveying profession and for public agencies.

The GAF can be in the form 0.999xxxxx or the reciprocal value of 1.00xxxxxx.

(1/0.999xxxxx = 1.00xxxxxx)

Which ever form is used, it is imperative that it is applied correctly when converting between "Grid to Ground" and "Ground to Grid".

For the purposes of this book, the GAF will refer to the reciprocal form of 1.00xxxxxx.

 Eight (8) decimal places are sufficient enough for most surveying projects.

There are two methods that can be used to apply the GAF. Method 1 is where the origin is located at N=0, E=0 and Method 2 where the origin is located somewhere other than N=0, E=0.

Method 1 ~ Origin (N=0, E=0)

Ground Coordinate = Grid Coordinate * GAF

Method 2 ~ Origin (N1, E1)

KML & KMZ

Ground Northing = ((Grid Northing - N1) * GAF) + N1
and
Ground Easting = ((Grid Easting - E1) * GAF) + E1

Method 1 is the preferred method. It is simple to use and to visually distinguish between Ground and Grid coordinates.

 **Method 2** is used for holding NGS Grid values so that Grid and Ground values are very close to being the same. [Note: Some municipalities will also require northing and easting translations and/or rotations]

The GAF can be derived by utilizing GPS software such as Trimble Business Center, as shown on an NGS Data Sheet or as defined by a private/public agency.

For most locations within a state plane zone, the GAF will be over 1.00xx meaning that the the ground coordinate will be larger than the grid coordinate after the transformation when using Method 1 above.

Another way to look at it is that the grid plane is below the surface for most of the state plane zone. Out on the fringes of the zone, the grid plane can be above the surface especially at lower elevations.

KML & KMZ

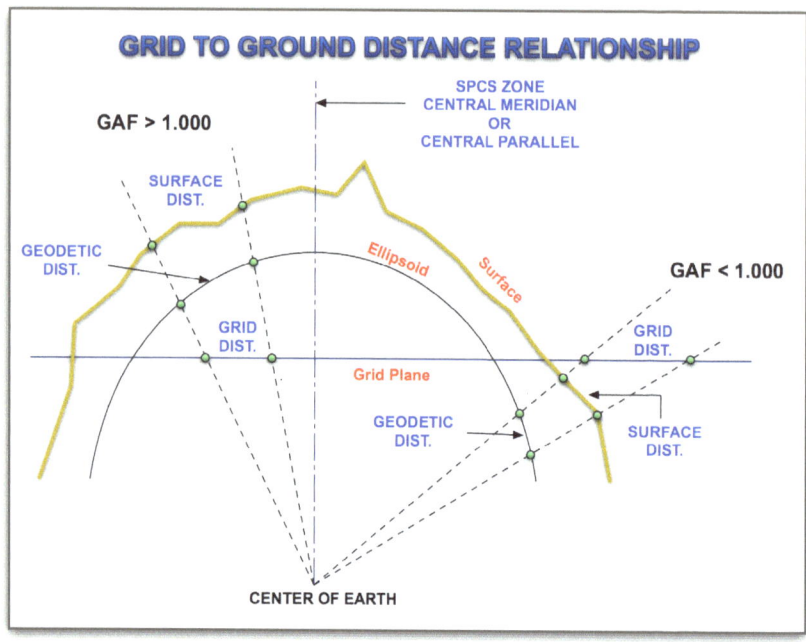

 For more information on converting Grid to Ground coordinates see "Coordinate Transformation" - Book 9" and "What was that Formula - Book 11" of this book series.

http://www.cc4w.net/ebooks.html

Below are the three zones for State of Arizona. The areas in blue are where the grid plane is above the ellipsoid and at low elevations are above the surface. In these areas, the surface distance will be shorter than the grid distance.

KML & KMZ

For the purposes of this book, the GAF will be given. The above information is for your general information and knowledge.

KML & KMZ

REPRESENTATIVE PARCEL

This Representative Parcel will be utilized for all four platforms (Traverse PC, MicroStation, Civil 3D & Trimble Business Center) to create a basic KML/KMZ file to import into Google Earth.

For the purposes of this book and the step-by-step processes, the following will be provided:

- Parcel Legal Description
- Ground Coordinates for the parcel boundary.
- Grid Adjustment Factor
- State Plane Zone

It is assumed that you know how to setup a project in at least one of the four platforms, import the ground coordinates and generate linework for the boundary parcel. We will show you how to create a KML/KMZ file for the parcel that is geo-referenced for Google Earth.

To give you a sneak peek, below is were we are headed with the step-by-step instructions in this book:

KMZ file loaded into Google Earth.

Parcel Legal Description

Those portions of the Southwest quarter of Section 25, the Southeast quarter of Section 26 and the Northeast quarter of Section 35, Township 41 North, Range 30 East of the Gila and Salt River Meridian, Apache County, Arizona, described as follows:

Commencing at a found 3 1/2 inch Bureau of Land Management (BLM) Department of Interior 2006 brass cap marking the common corner for Sections 25, 26, 35 and 36 of said Township, from which the quarter corner for Sections 26 and 35 of said Township bears South 89°21'45"

West, 2638.40 feet being marked with a 3 1/2 inch BLM 2006 brass cap;

Thence South 89°21'45" West, 305.47 feet along the south line of said Section 26 to a point on the northwesterly right of way line of route U.S. 160 (Tuba City - Four Corners) being the **POINT OF BEGINNING**;

Thence South 34°09'26" West, 47.03 feet along said northwesterly right of way line;

Thence North 55°52'03" West, 599.82 feet;

Thence North 34°09'22" East 600.24 feet;

Thence South 55°49'38" East 596.45 feet to a point that lies on said northwesterly right of way line being the point of curvature of a non-tangent circular curve to the right having a radius of 7539.44 feet;

Thence along said northwesterly right of way line, from a local tangent bearing of South 32°26'34" West along said curve a distance of 225.60 feet through a central angle of 1°42'52" to the point of tangency;

Thence South 34°09'26" West, 327.23 feet continuing along said northwesterly right of way line to the **POINT OF BEGINNING**.

Said parcel of land containing 359,655 square feet (8.257 acres), more or less.

KML & KMZ

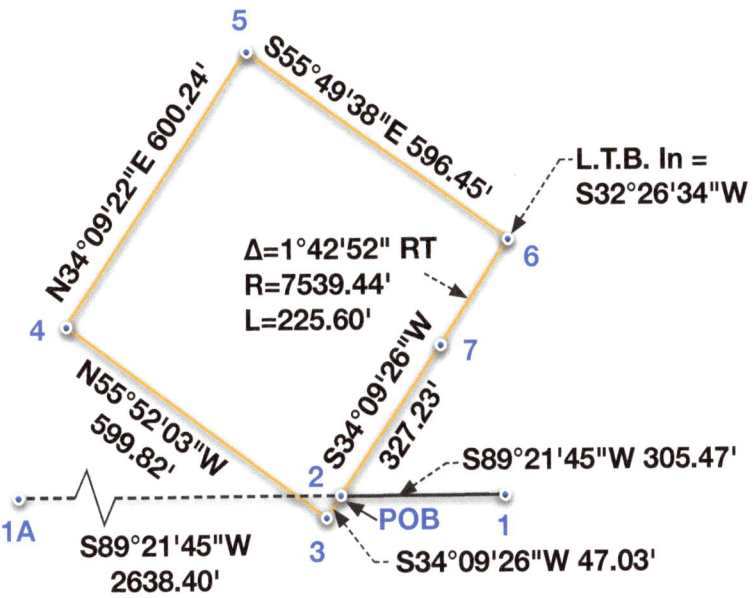

Ground Coordinates		
Pt #	Northing	Easting
1	2157691.70913	1016642.48026
2 (POB)	2157688.31036	1016337.02636
3	2157649.39711	1016310.62333
4	2157985.95806	1015814.13046
5	2158482.66135	1016151.13398
6	2158147.63943	1016644.60747
7	2157959.09703	1016520.75779
8 (RP)	2162192.24047	1010281.87551
1A	2157662.35359	1014004.24314

 The ground coordinates are directly related to the Grid Coordinates via the GAF. See the GAF information shown earlier in this book. They are not local (random) coordinates. Local coordinates will not geo-reference to Google Earth by the methods shown in this book. Converting local coordinates are outside the scope of this book.

Grid Adjustment Factor

The Grid Adjustment Factor (GAF) a.k.a. combined factor, for this project is 1.00024915.

State Plane Zone

The State Plane Geographical Coordinate System is as follows:

NAD83 Arizona State Plane, East Zone, International Foot

KML & KMZ

TRAVERSE PC

The following are the steps needed to create a KMZ file using Traverse PC and open it in Google Earth.

- Create a TPC File and Select the Coordinate Reference System
- Import Ground Coordinates and Convert to Grid
- Review Imported Points in the Points Manager
- Create a Drawing with the Boundary and Other Features
- Export the Drawing to a KMZ File
- Preview the Exported KMZ File in Google Earth

Create a TPC File and Select the Coordinate Reference System

In Traverse PC, a survey is assigned a Coordinate Reference System (CRS) that transforms grid coordinates to geodetic positions and vice versa. The CRS we select will transform the coordinates we import in the next section into the geodetic positions TPC exports via the KMZ file.

1. Create a new TPC file and set the Coordinate Reference System as shown here.

KML & KMZ

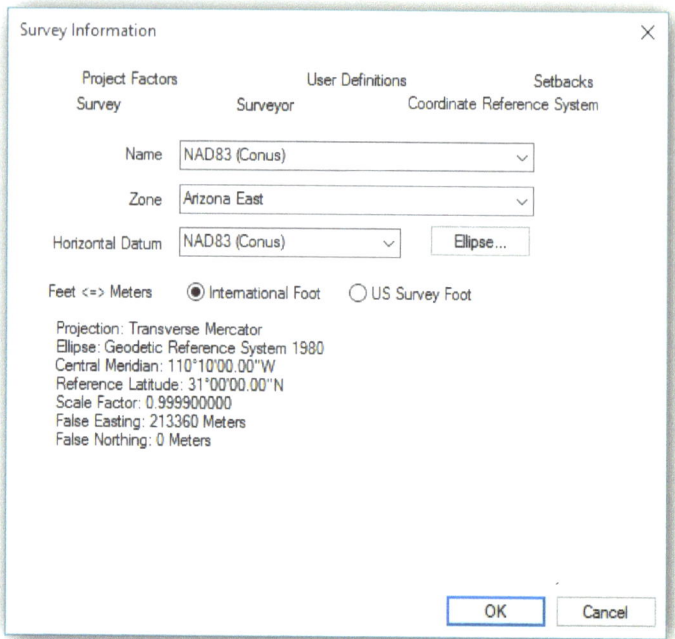

2. Choose OK and close the New Survey Hints dialog.

Import Ground Coordinates and Convert to Grid

The coordinates we will be importing are ground coordinates with all elevations set to 5200. TPC works in grid coordinates so we will need to convert ground to grid when we import them.

Note: You can enter the points manually in the Traverse View or Points Manager if you don't have a file to import. You could also wait to convert the coordinates later. The method we are using here simplifies the process.

1. From the TPC Desktop, choose File | Import.

KML & KMZ

2. From the Type pulldown, select the ASCII format.
3. Click the Settings button to the right (it looks like a gear) to open the ASCII Settings dialog.
4. Turn on (X) Convert Coordinates and click Settings.
5. Set the options as shown here with the GAF set to 0.999750912. (*1 / 1.00024915*)

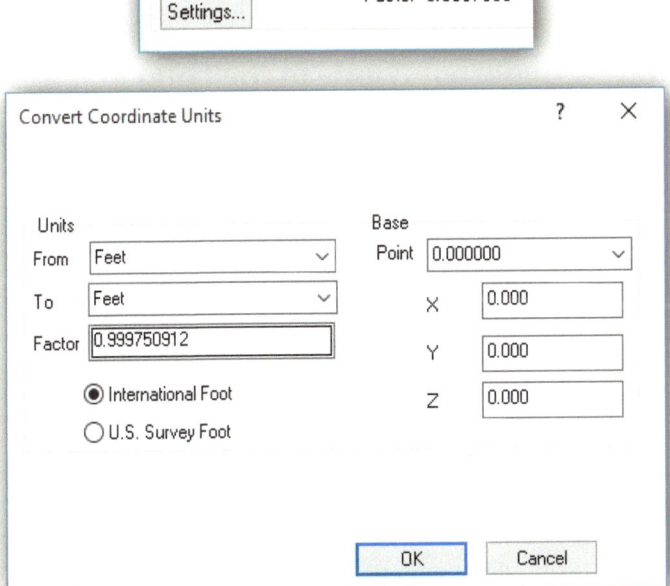

KML & KMZ

 **Note:** Traverse PC will **multiply** the coordinates by the GAF. The coordinates on the right are the base coordinates that the Factor will be applied from. If we enter 5200 in Z, the existing elevations of our points (5200) will not change.

1. Choose OK and OK the ASCII Settings dialog.

2. Choose Import.

The Ground coordinates are converted to Grid coordinates for us on import.

Review Imported Points in the Point Manager

Our data is now set on the NAD83 Arizona East grid so let's check the results in the Points Manager.

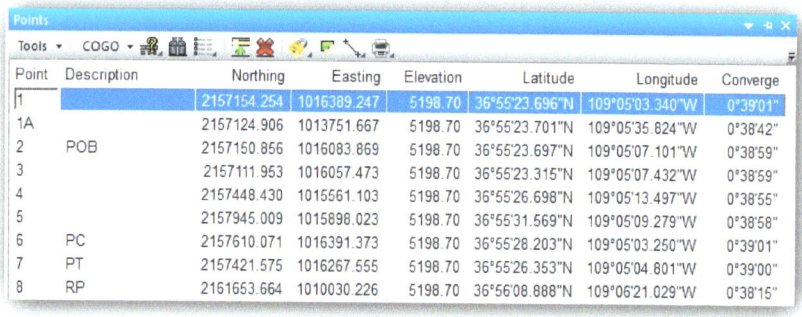

Notice that all of the coordinates (including the Elevation) have been reduced based upon the GAF we used to Convert Coordinate Units.

KML & KMZ

 **Note**: The Points Manager can be formatted to display various information about the points. Choose Tools | View | Format View.

Create a Drawing with the Boundary and Other Features

Our next step is to create the drawing that we ultimately want to see in Google Earth. In TPC, we do this by recalling the imported points into traverses using the point sequence that defines the feature we want. TPC creates the drawing directly from the traverses and points based on how we tell the program to display them.

1. **Create a BOUNDARY traverse** using the Deed w/ curves Traverse View Format and the Property Lines Traverse Drawing Settings.

2. Recall points 2 through 7, Enter the Radius 7537.56 on point 7 and recall point 2 to close the traverse.

3. **Create a RADIUS POINT traverse** using the Deed Traverse View Format and the Property Lines Traverse Drawing Settings.

4. Recall points 6, 8, 7 to show the radius point and close the RADIUS POINT traverse.

5. **Create a CONTROL traverse** using the Deed Traverse View Format and the Property Lines Traverse Drawing Settings.

6. Recall points 1 and 1A and close the CONTROL traverse.

KML & KMZ

7. Modify the Traverse Drawing Settings for each of the three traverses just created to use different line types, line widths, colors and point symbols.

Note: We set the BOUNDARY traverse to Green with .03 weight Solid lines and 3/8" IRON PIN FOUND symbol. We set the CONTROL traverse to Blue with .01 weight Solid lines and the Section Corner symbol. We set the RADIUS POINT traverse to Red with .00 weight Solid lines and the Rebar symbol.

Once we have recalled the points into our traverses, we make sure the traverses are tagged (checked) so they will be displayed in the Drawing View. This drawing is what we will export to a KMZ file. Once the drawing looks right, it will be right in Google Earth.

8. Click the Drawing View button on the TPC Desktop Navigation Toolbar to open the New Drawing dialog box.

9. Choose the Empty drawing template from the pull-down list of templates at the bottom and choose OK.

10. Go to the Traverses Manager and tag (check) all of the traverses except the original coordinates traverse (if you have one).

Note: You won't have an original coordinate traverse if you entered the points directly in the Points Manager.

KML & KMZ

11. Go to the Drawing View and choose Tools | Print | Page Setup to open the Page Setup dialog box.

12. Choose ARCH D as the Paper Size, Landscape as the Orientation, set all of the Margins to 0.5 and choose OK.

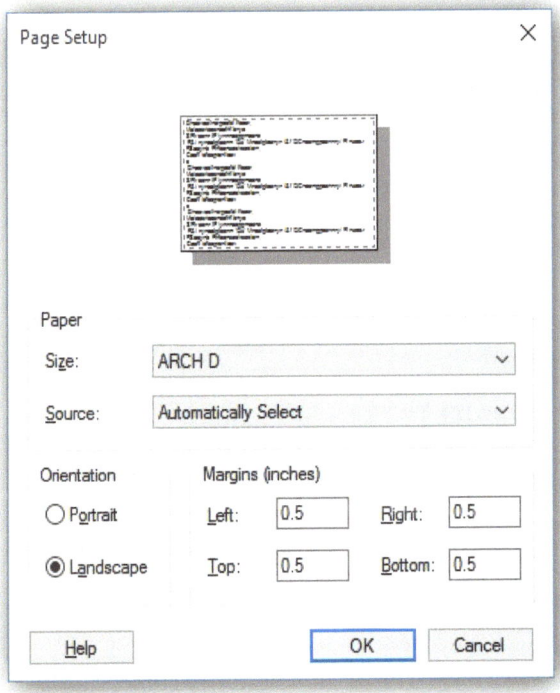

13. Zoom Extents on the drawing and it should look something like this:

KML & KMZ

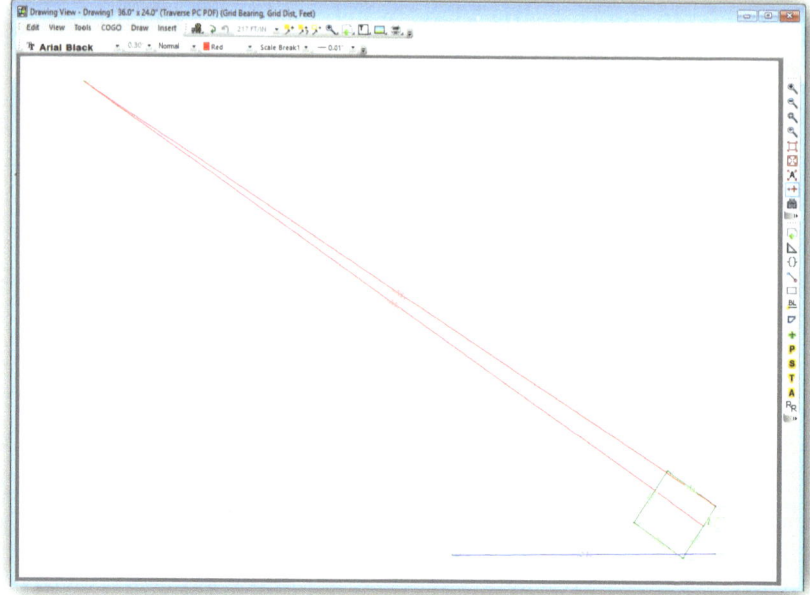

Export the Drawing to a KMZ File

Let's send it out to Google Earth and see how it looks.

1. Go to the File menu and choose Export.

2. Answer Yes when you are asked if you want to save the file.

3. Set the Type to Google KMZ.

 Note: You can export either KML or KMZ format files. The file size difference in this example from Traverse PC is 36.3 KB for the KML as opposed to 2.82 KB for the KMZ file.

4. Click the Settings button to the right (it looks like a gear).

KML & KMZ

5. Set the KMZ settings as shown here and choose OK:

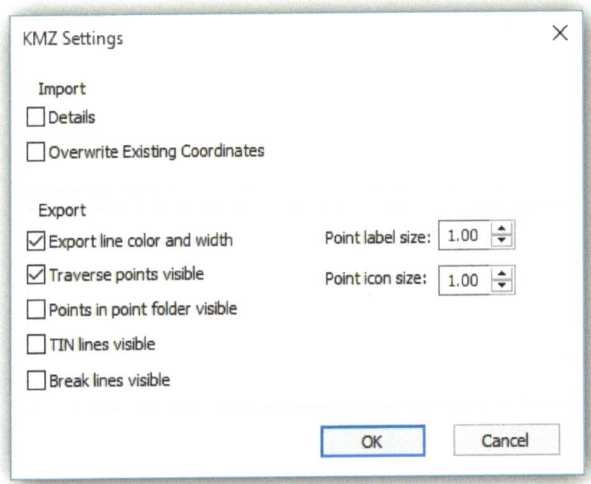

 **Note**: The sizes you set are an arbitrary value. As you zoom in and out in Google Earth, they will remain a constant size so the goal here is to provide a readable size that works with the data you are exporting.

6. Set the What List to Entire Drawing: Drawing1 and choose Export.

KML & KMZ

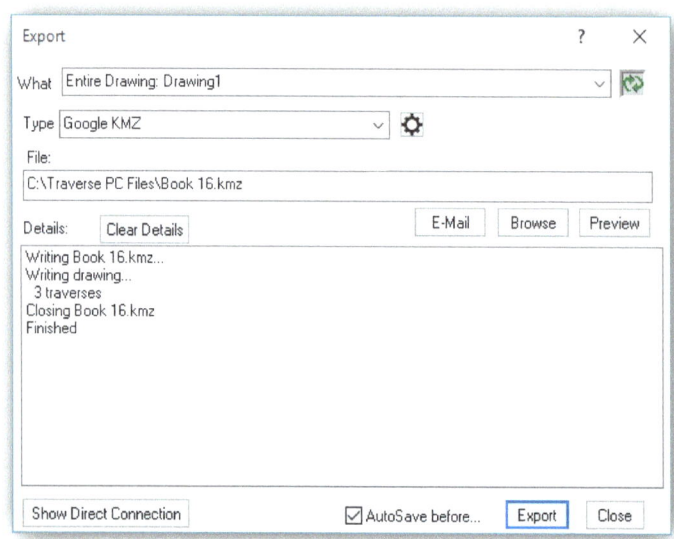

Preview the Exported KMZ File in Google Earth

1. Click the Preview button in the Export dialog box and as long as Google Earth is installed, the KMZ file will open and fly you in to the location.

Note: Google Earth should have associated KML and KMZ files with itself in Windows when it was installed. If the file doesn't open in Google Earth when you click the Preview button, you may need to tell Windows what program to open it with.

KML & KMZ

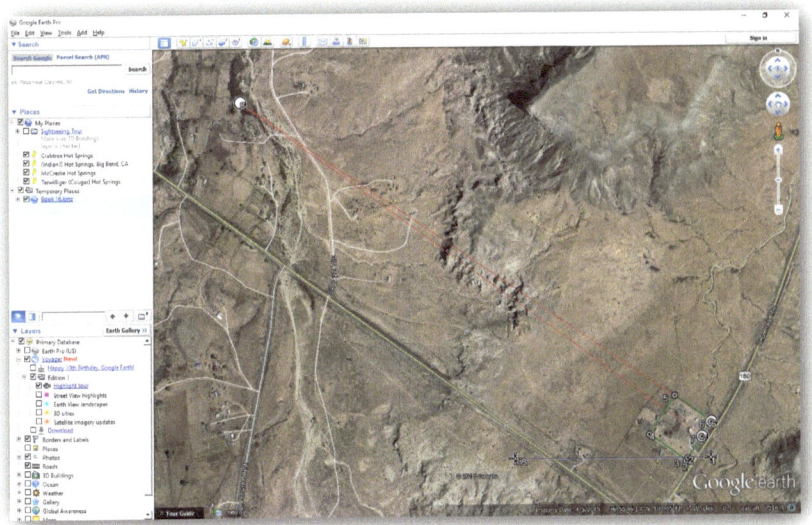

2. Zoom in on the property boundary and click on the point label or point symbol for point 7.

KML & KMZ

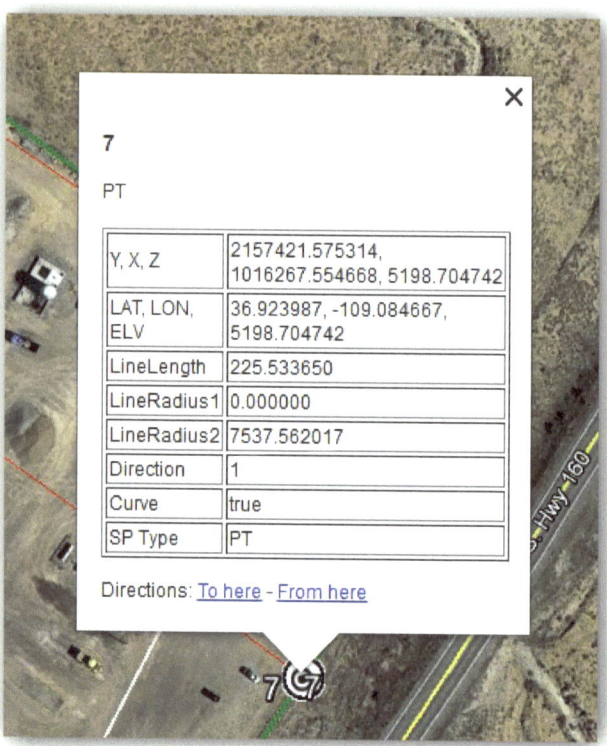

The point information balloon displays the Point Number, Point Description, Grid Coordinates, Geodetic Position, Grid Line Length, Curve Radius (grid) and Point Type.

Note: Traverse PC includes a full range of Google Earth compatible point symbols so you can use whatever symbols you choose and have them show up in Google Earth.

Clicking on the file name under Temporary Places in the Places panel on the left displays the program and program version that created the file.

3. Expand the file under Temporary Places by clicking the plus sign (+) next to the file name.

4. Expand Traverses by clicking its plus sign (+), expand the BOUNDARY traverse and expand the Points.

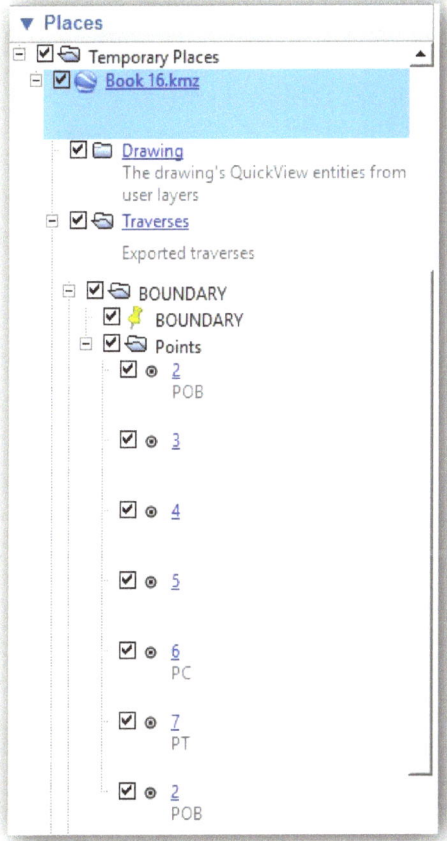

You can turn anything on or off by tagging (checking) or untagging (unchecking) it in the list. You can click on any object in the list to see its information balloon.

MICROSTATION

The following are the steps needed to create a KMZ file using MicroStation and open it in Google Earth.

- Create a new project DGN file
- Add the ground coordinate points and linework
- Create a new DGN file for NGS Geographical Coordinates
- Select the Geographical Coordinate System
- Attach the project DGN file
- Scale the project DGN file by the GAF
- Create a KMZ file
- Open the KMZ in Google Earth

Start by creating a new project DGN file. Pick a name that will assist you in remembering what the DGN file is for. I named the example DGN file "Book 16.dgn" for this session.

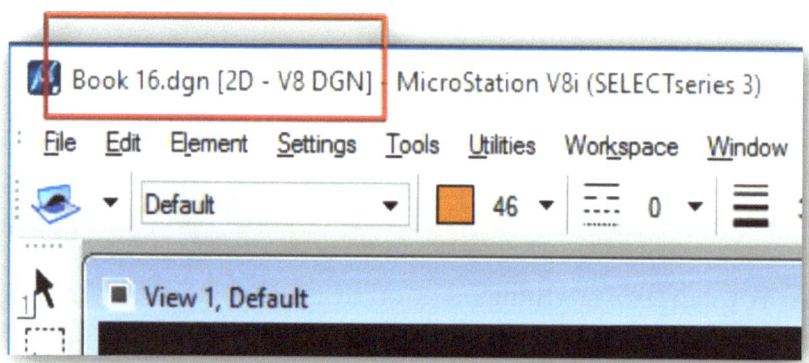

Next you will need to import the ground coordinate points of the Representative Parcel into this DGN file.

KML & KMZ

 **Note**: For this session, ground coordinates will be utilized therefore all points, linework, etc. will be in ground values. This makes it useful if you are plotting information from legal descriptions, record drawings, etc. which are in ground values. There is no conversion necessary using the method described in this session.

There are several ways to import the ground coordinate points into a DGN file. There are several programs such as InRoads and SimPro that will interface with MicroStation in order to import coordinate points.

For this session, we are going to use MicroStation's built in commands to place the points. There is no need for additional software to perform this operation.

MicroStation has a cool tool called "Place Note" that will let you dynamically plot the ground coordinate point and name it at the same time.

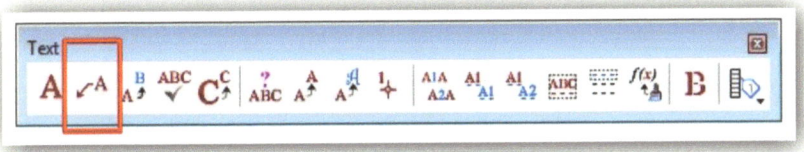

There are several settings associated with this command. You will need to experiment with these settings to see the effect that they have.

KML & KMZ

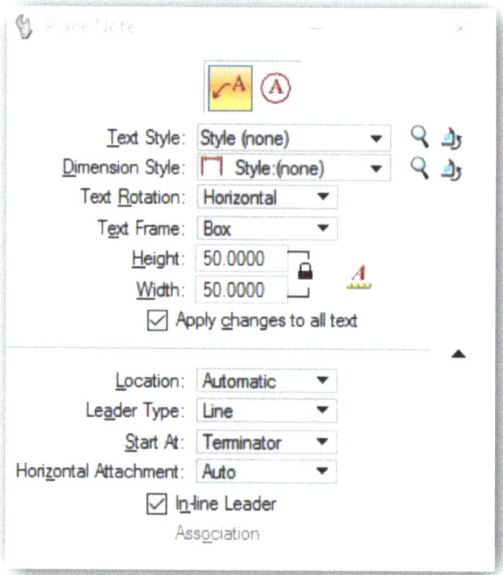

Set the Height and Width to a value that you can read the text when you "Fit View" to view all of the points.

Select the Level, Color, Linestyle and Weight.

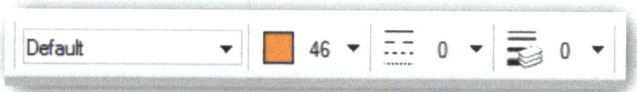

Open the "Place Note" and "Key-in" commands.

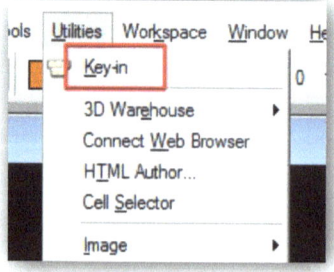

With "Place Note" and "Key-in" running, position the "Text Editor" dialog box close to the "Key-in" dialog box.

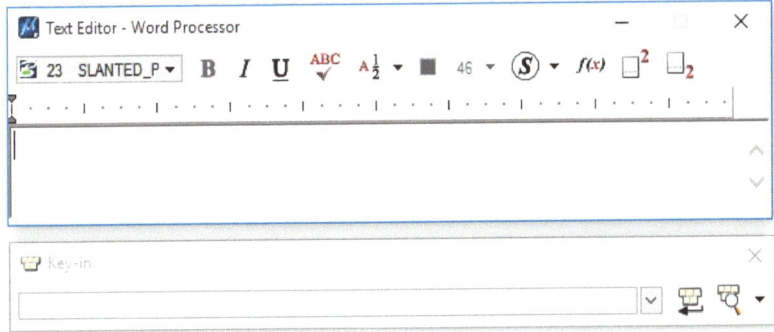

In the "Text Editor" type "1" for the first ground coordinate point number.

In the "Key-in" type the follow command:

xy=1016642.48026,2157691.70913

 **Note:** With the "xy" command you need to enter the Easting [comma] Northing in that order.

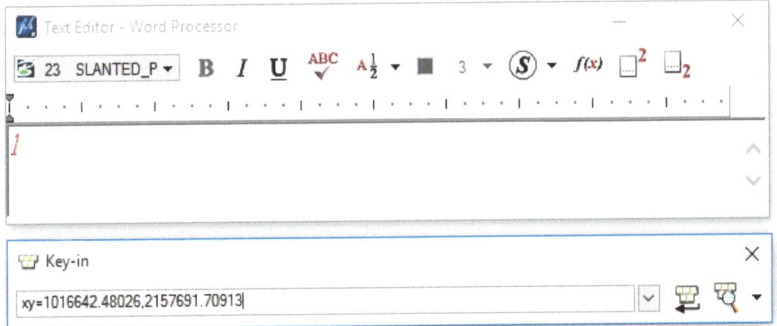

KML & KMZ

Hit enter to place the point.

The tip of the arrow will be at the ground coordinate point location. The text is dynamic. Move your mouse to place the text in a position that is clear of any linework or other points.

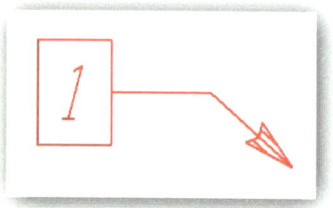

Repeat the above process for each of the ground coordinate points.

When all ground coordinate points have been entered, do a "Fit Views" to see all of the points on the screen.

Your screen should look similar to the following:

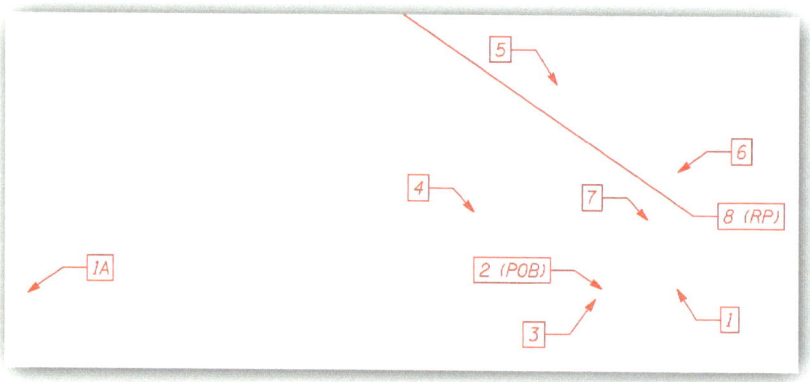

Change the color to "Green" for the parcel boundary linework.

KML & KMZ

Start the "Place Smartline" command.

Zoom in to the parcel area. Use the snap tools to connect the straight line segments of the parcel starting at point 7 then clockwise around the parcel to point 6.

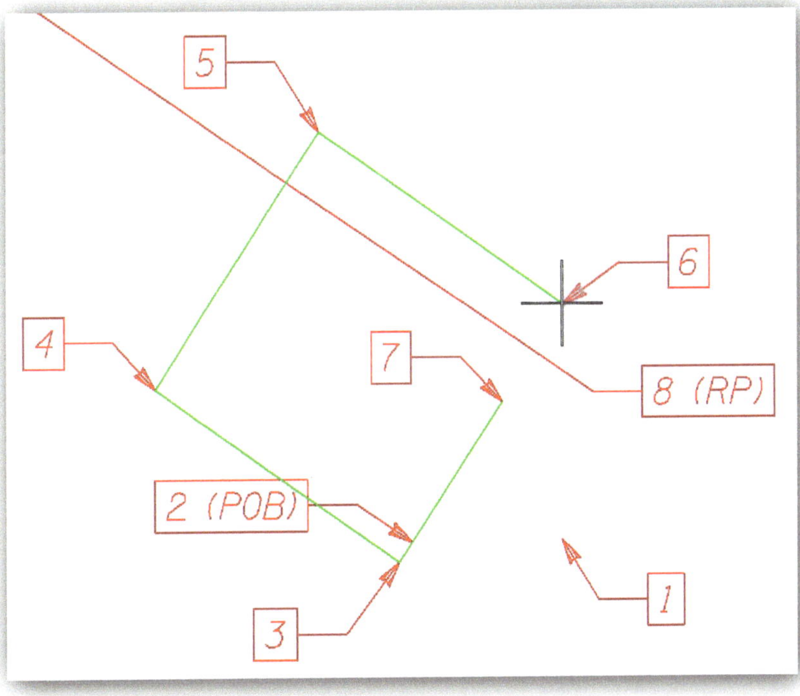

KML & KMZ

Use the "Place Arc" command to draw the curved segment between point 6 and 7.

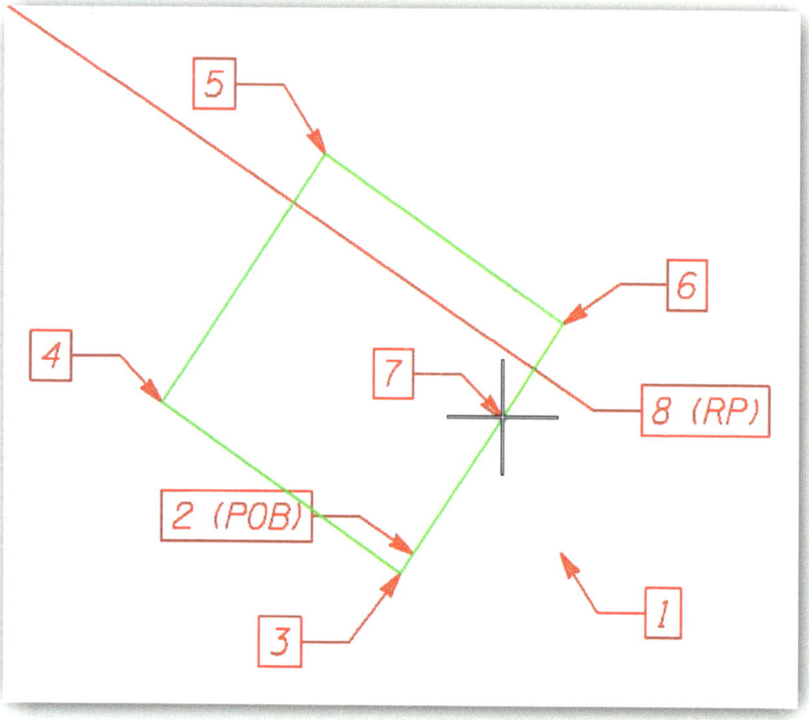

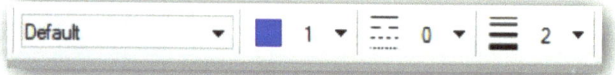

Change the color to "Blue" for the section linework.

Draw a line from point 1 to point 1A for the section line.

KML & KMZ

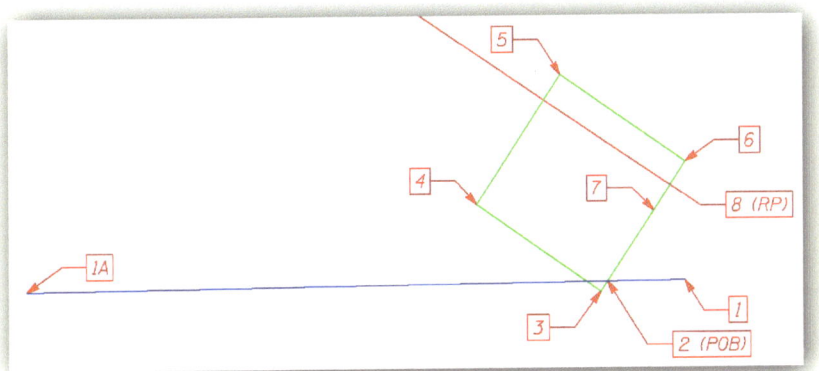

Now that the points have been plotted and the linework added to the DGN file, we are ready for the next step.

Create a new DGN file and name it "NGS Grid Data-GC.dgn" or some other name that indicates that it is for Grid values.

In this dgn file, select the "Geographic Coordinate System" from the "Tools" drop down menu.

KML & KMZ

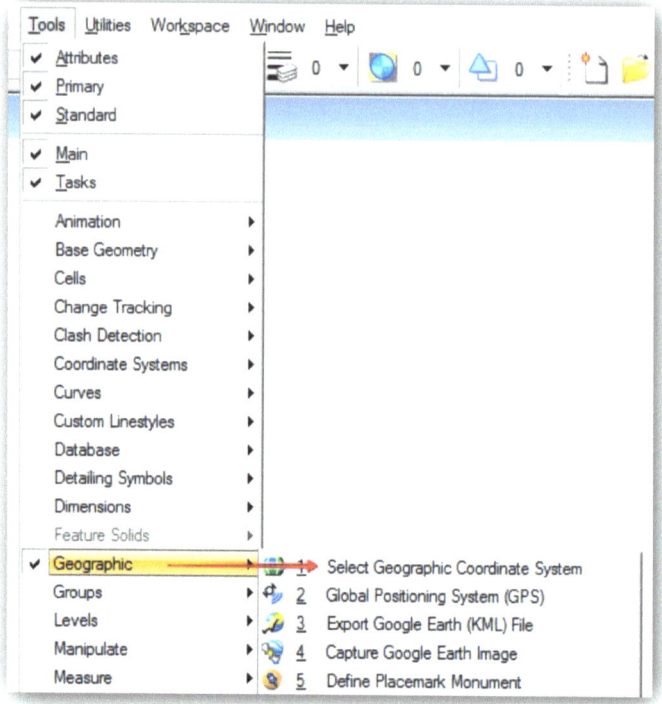

Click on the "From Library" icon to open the library.

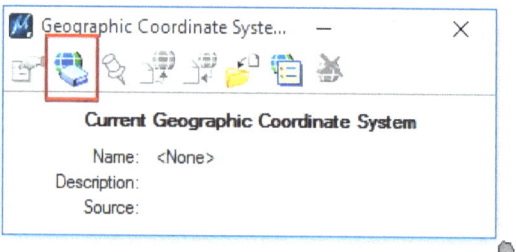

In the Library folder, open Projected (northing, easting, ...)

-> North America

-> United States of America

-> Arizona.

KML & KMZ

Select AZ83-EIF - NAD83 Arizona State Planes, East Zone, International Foot.

Click "OK".

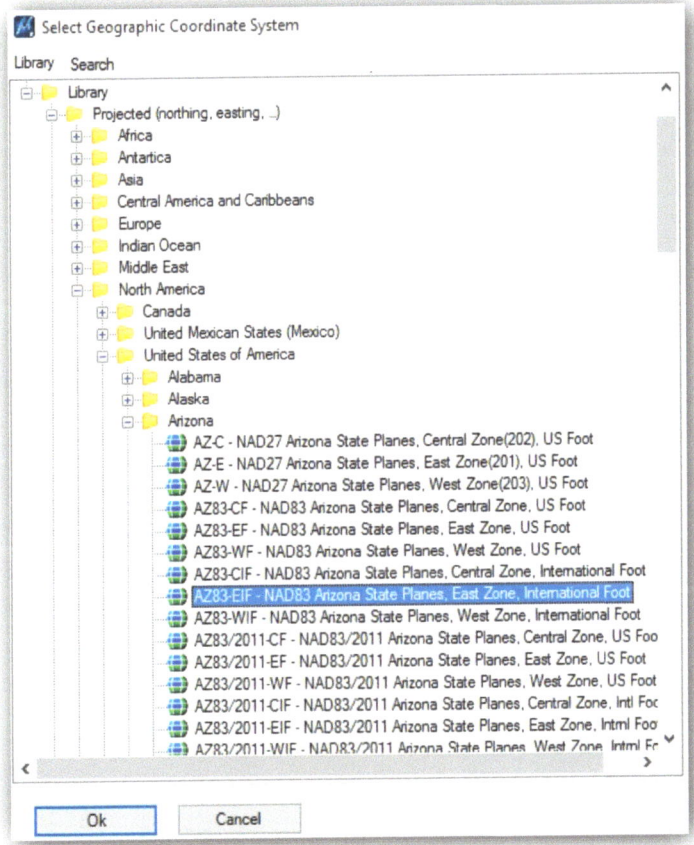

You now have the Geographical Coordinate System set for this parcel in the DGN file.

KML & KMZ

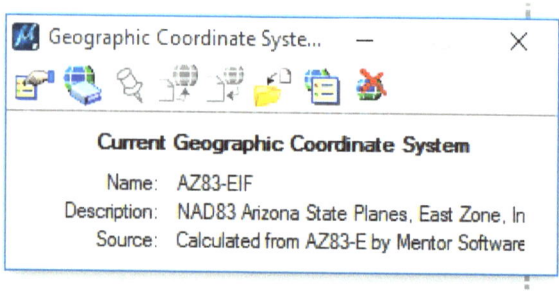

Now open the Reference Dialog Box.

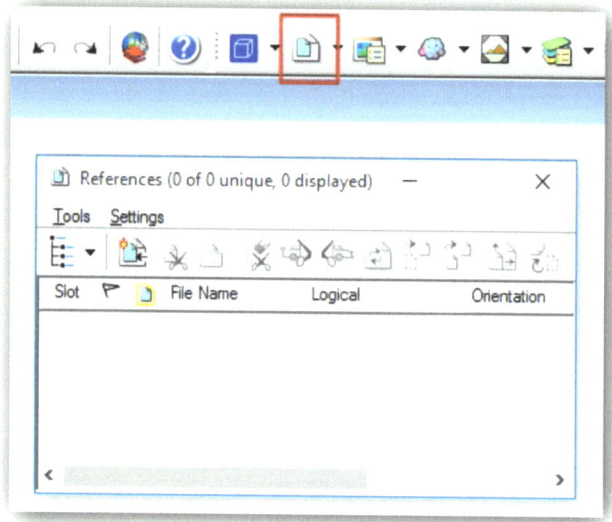

In the Reference Dialog Box, attach the "Book 16.dgn" file using the "Coincident" attachment method.

KML & KMZ

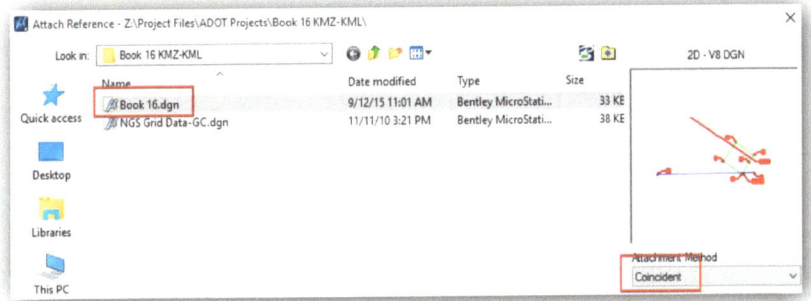

Click "open" to attach this file.

With "Book 16.dgn" selected in the "Reference Dialog Box" input the GAF of 1.00024915 for the parcel and type enter.

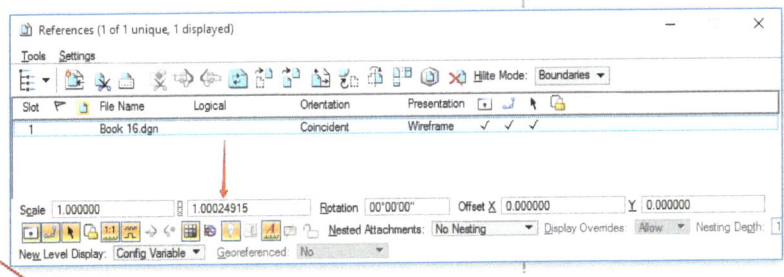

What this command is doing is scaling the reference linework from ground to grid at the origin of N=0, E=0. The linework in the "Book 16.dgn" is still at ground values. By using the referencing scheme, you perform your work in the ground file and then when you are ready, you open the NGS dgn file to create the KMZ file in grid.

Click "Fit Views" to fit the scaled ground to grid linework to the screen.

KML & KMZ

You should see something that looks like the following:

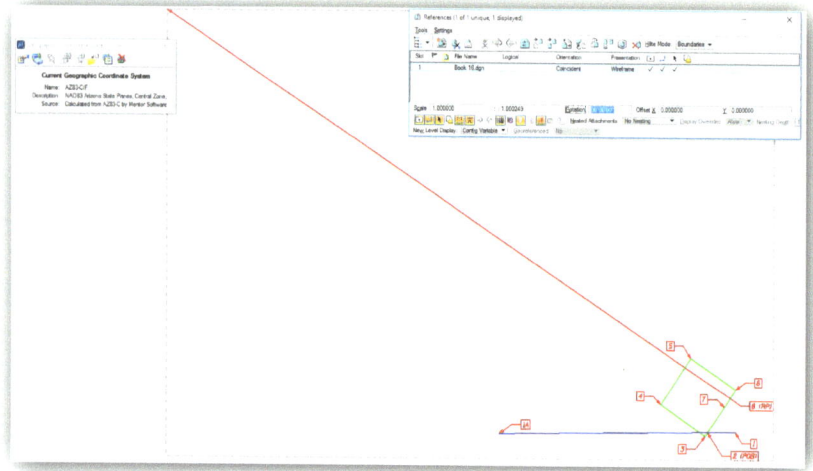

We are almost there.

To create the KML file click on the "Export Google Earth (KML) File" command.

KML & KMZ

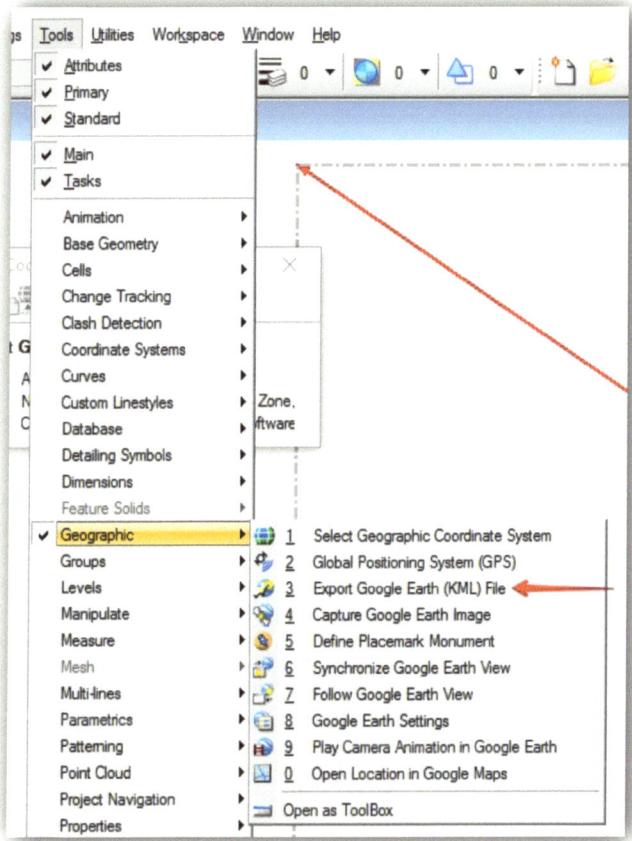

A save dialog box will open. Type in the name of the file, select the format (KML or KMZ) and click "Save". KMZ is a much smaller file which is easier to share with others via email especially if you have a lot of elements in your DGN file.

KML & KMZ

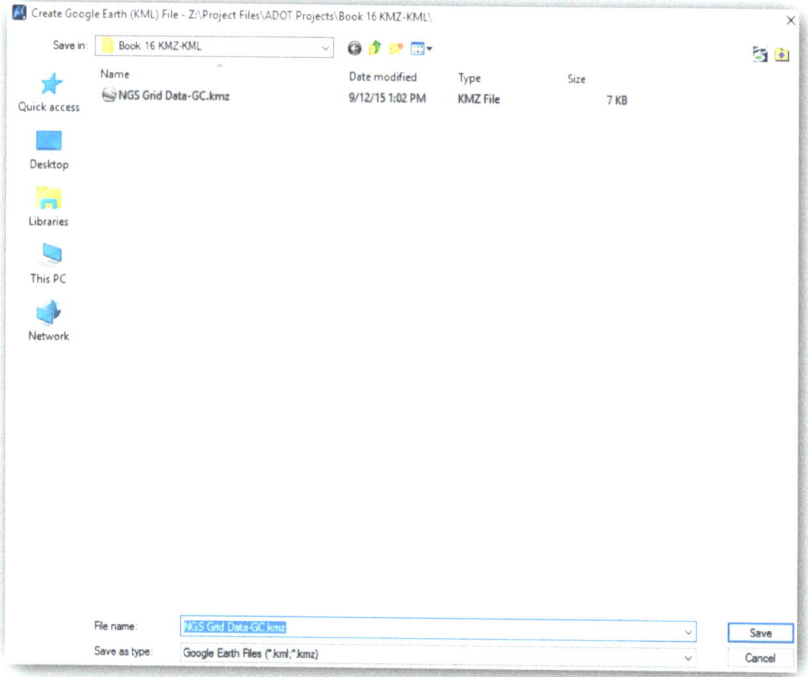

Google Earth should open automatically. If not, use File Explorer to navigate to where you saved the file and double click on the "NGS Grid Data-GC.kmz" file.

If everything was entered correctly, you should see the following:

KML & KMZ

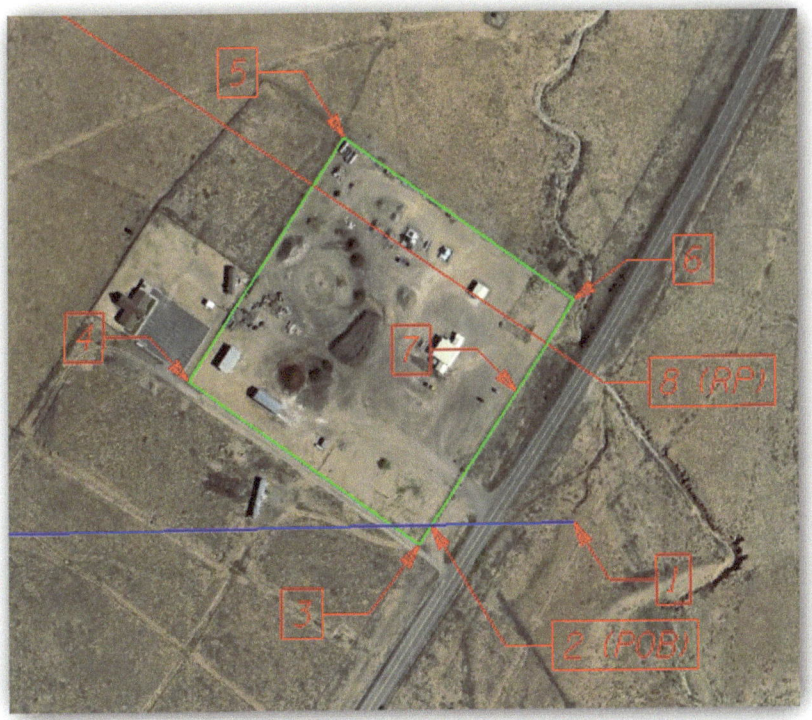

At this point, return to the "Book 16.dgn" file and add text, shapes, linework, move things around as desired. Then open the "NGS Grid Data-GC.dgn" file, recreate the KMZ file and open in Google Earth.

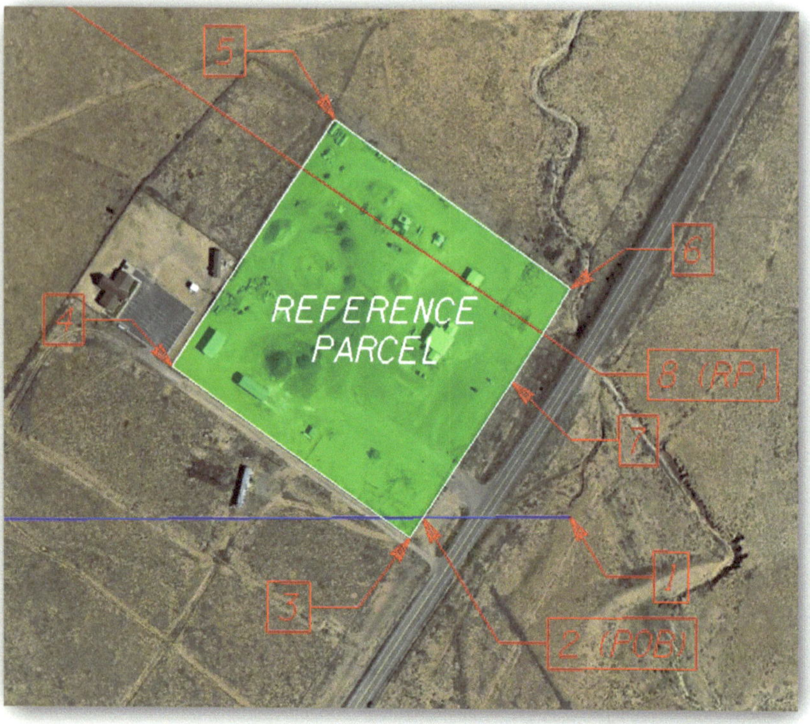

Whatever you draw in MicroStation, can be imported into Google Earth through a KMZ file.

I often work on large projects with 100's of parcels. It is real helpful to create different colored shapes for numerous parcels and bring them into Google Earth for viewing and analyzing.

The following is a sample of a large project near Camp Verde, Arizona.

KML & KMZ

Below is the linework in MicroStation:

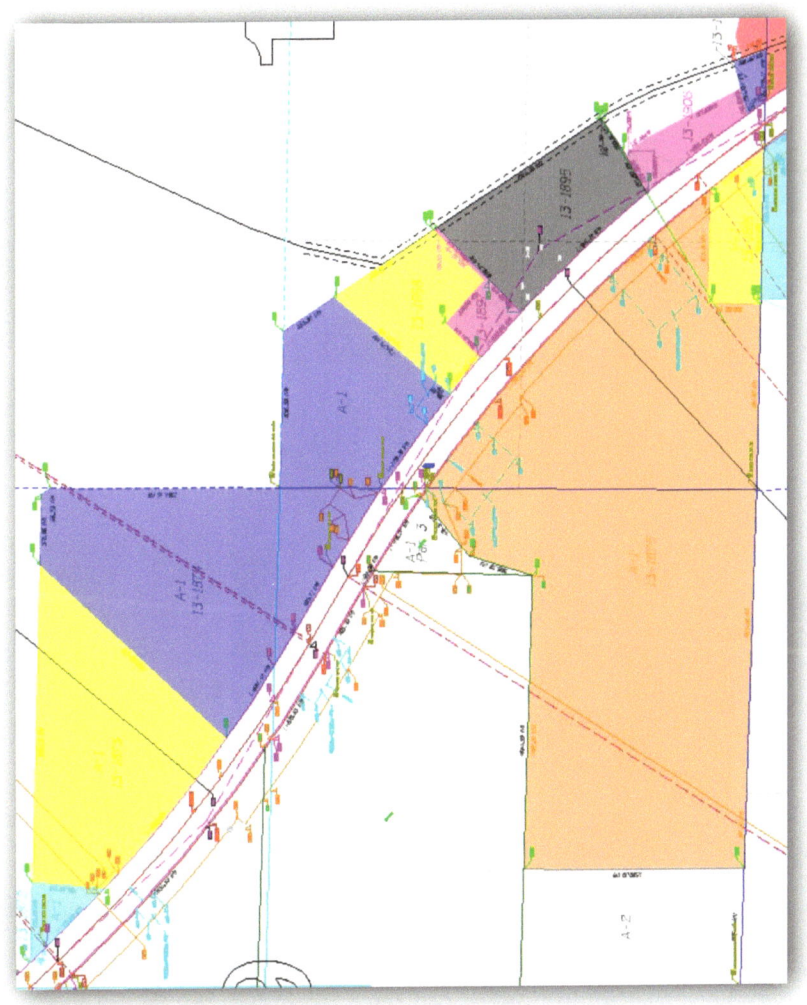

KML & KMZ

Below is the Google Earth view:

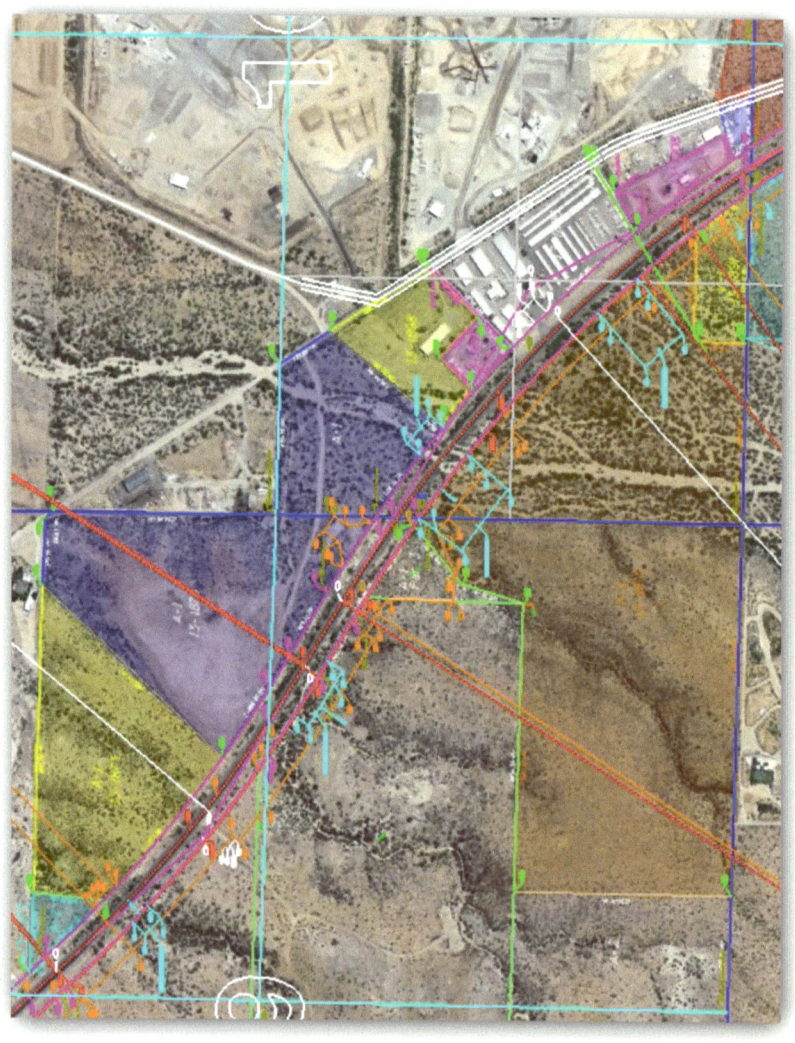

CIVIL 3D

The following are the steps needed to create a KMZ file using Civil 3D and open it in Google Earth.

- Create a new drawing
- Use the existing "All Points" Point Group
- Edit Drawing Settings
- Select a Geographical Coordinate System
- Select Coordinate Method
- Create Line Work
- Label Lines
- Publish to Google Earth
- Save KMZ File
- View in Google Earth

Create a New Drawing

Create a new drawing in the customary manner. In this example, we will be reviewing the GE publishing procedure for Civil 3D 2012. (*Note: As of the writing of this book, Civil 3D 2013-2014 DOES NOT have the GE publishing feature*)

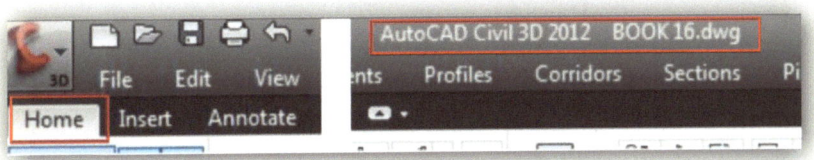

Create a Point Group

Utilizing the Representative Parcel Ground Coordinates, add them to the default _All Points_ group per the Civil 3D points procedure.

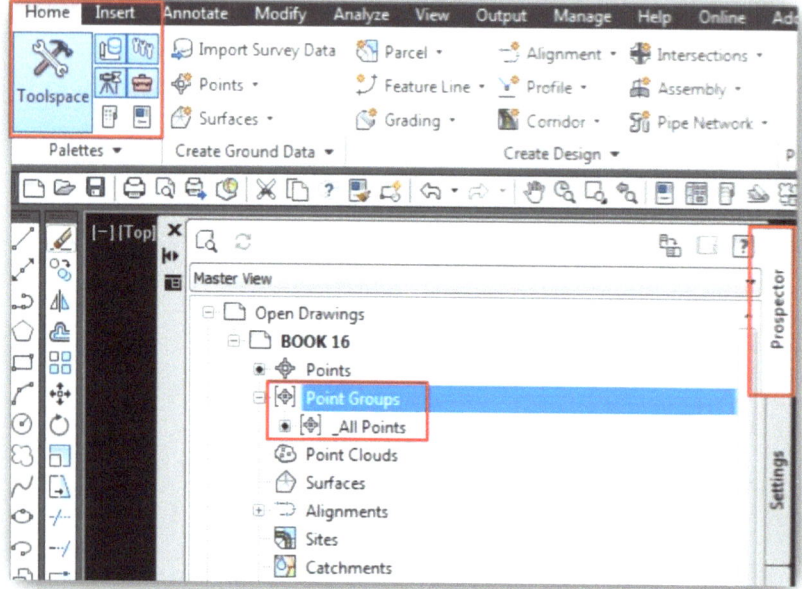

Edit Drawing Settings

It is possible to transfer a variety of spatial information contained in the drawing (_line work_) to GE such as parcels, alignments, surfaces. For this example, the Representative Parcel is a four-sided parcel.

GE will publish your drawing in the correct position on the aerial image only if we select the settings in the drawing that are consistent with the actual position of the drawing **on the ground**. To do this, we'll navigate to the Settings tab on the Toolspace menu for your (_active_) drawing **Book 16**.

KML & KMZ

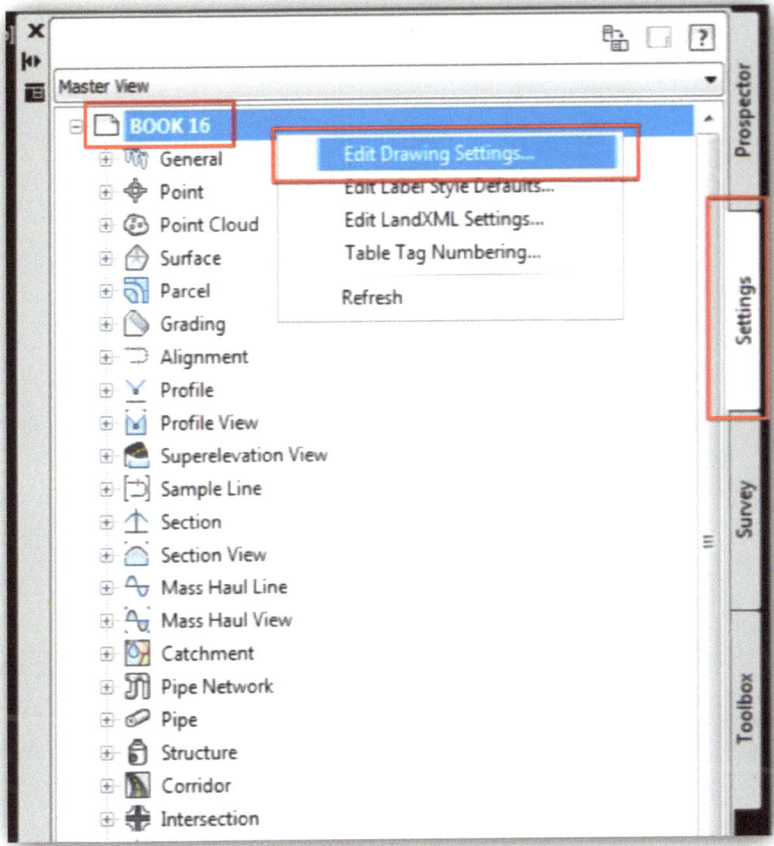

Select a Geographical Coordinate System

Pick the UNIT & ZONE tab. Select the coordinate system in which your drawing resides. For the Representative Parcel select Arizona State Plane, East Zone, International foot.

KML & KMZ

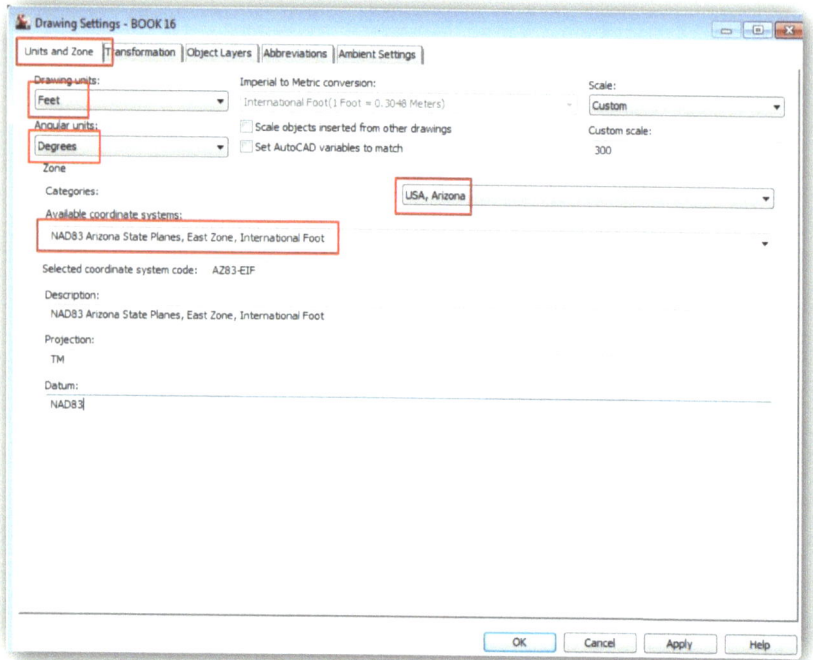

Most of us work with GROUND coordinates on our projects. GE works only with GRID coordinates and line work. We do not have to have grid coordinates or line work available for the Representative Parcel to be able to publish to GE direclty. Civil 3D can convert the line work to GRID when you publish to GE. To do this, check the "Apply Transformation Setting" box on the "Transformation Tab". This will apply the Representative Parcel GAF when publishing and move the line work "down to grid". Select <u>User-Defined</u> from the Computation pull down menu and enter the Representative Parcel GAF.

Set the scale factor as 0.99975091206.
(1 / 1.00024915)

KML & KMZ

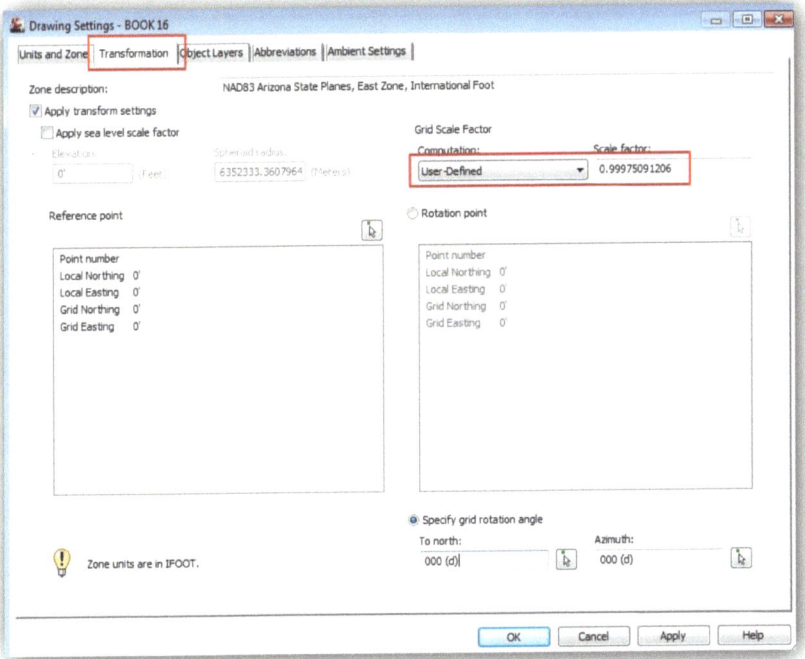

Select Coordinate Method

We have the ground coordinates for the corners of the Representative Parcel, so all we have to do now is get the ground coordinates for the corners (points) into **Book 16.dwg**. There are a couple of methods we can utilize to enter the coordinates into the drawing. In our example, we'll create points by Northing and Easting.

KML & KMZ

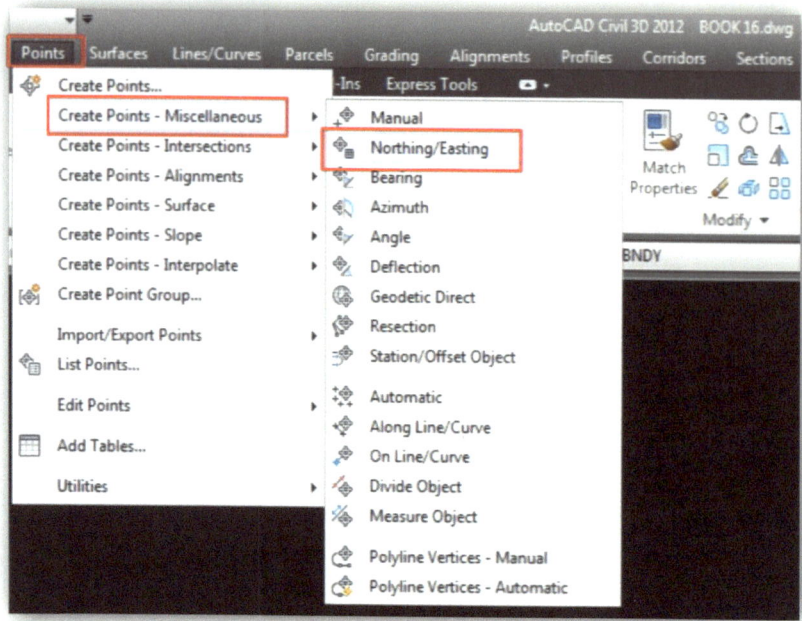

After all the points have been entered, select the scale in which you would like the points and line work to be displayed. This will be the scale that your drawing is published to and displayed in GE.

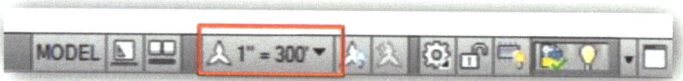

Create Line Work

Create line work for the Representative Parcel and the adjacent Section line, by drawing lines connecting the points.

KML & KMZ

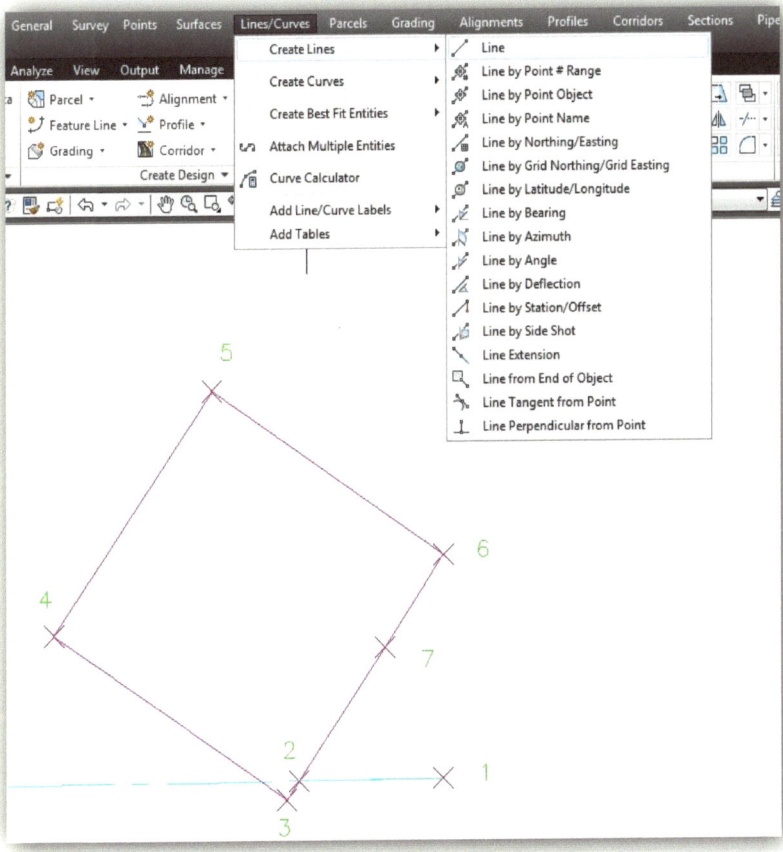

Label Lines

Label the line and curve information for the parcel and the Section line.

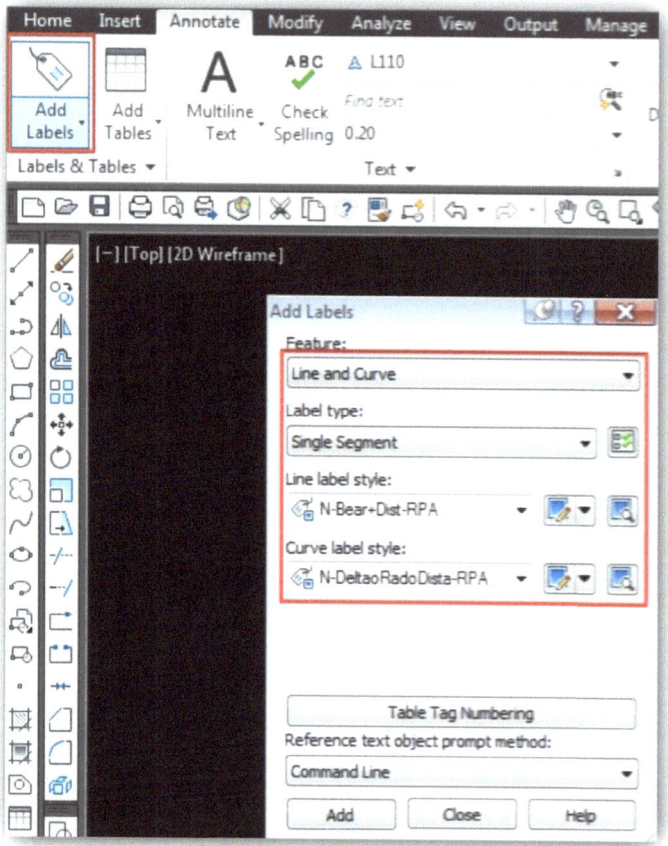

Careful placement of the labels will allow you to display the information in a relatively small area.

KML & KMZ

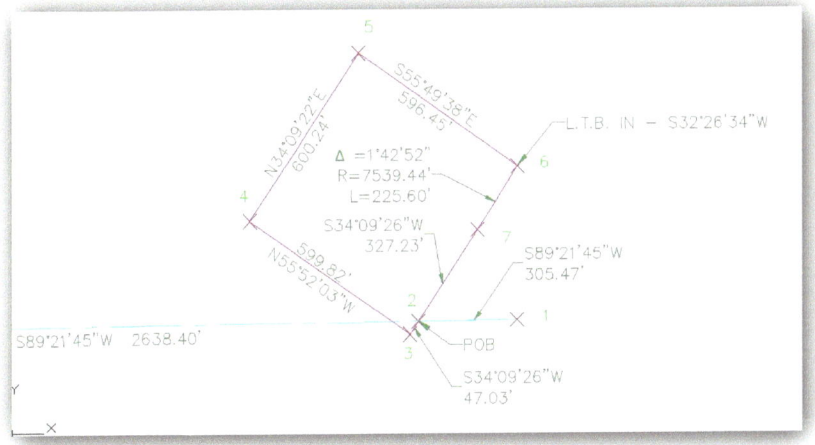

Publish to Google Earth

Now that we have the Representative Parcel drawn, we can publish to GE. Let's start by selecting the Output tab from the ribbon and then picking the "Publish to Google Earth" Command.

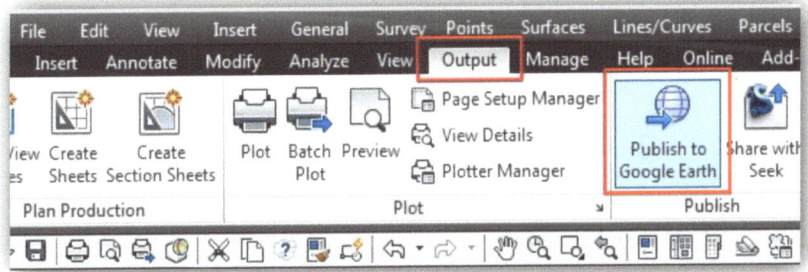

There are a few options to be selected and questions to be answered about the parcel (or display) and how you would like it to be viewed in GE. First is to the name the parcel.

KML & KMZ

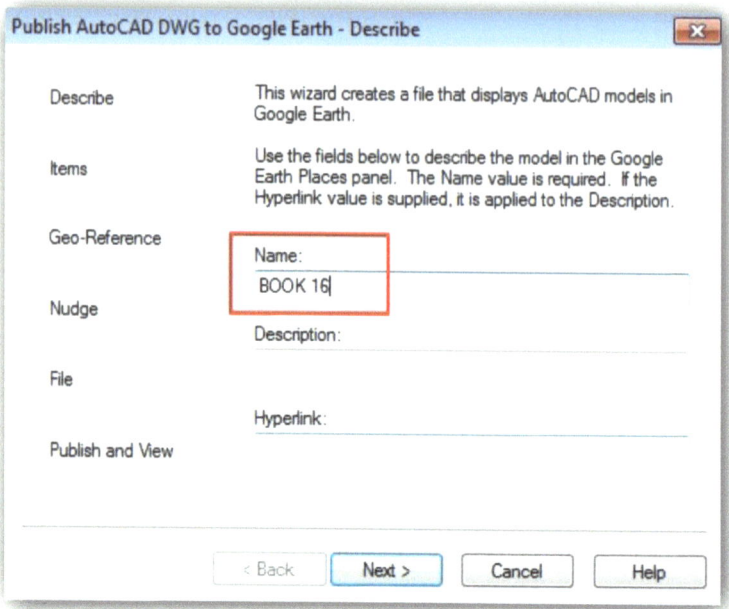

Next is to select what will be displayed.

We have the option to select "All model space entities" or only a portion of what is in the drawing by clicking "Selected model space entities" by clicking the "plus" button and choose the drawing entities.

KML & KMZ

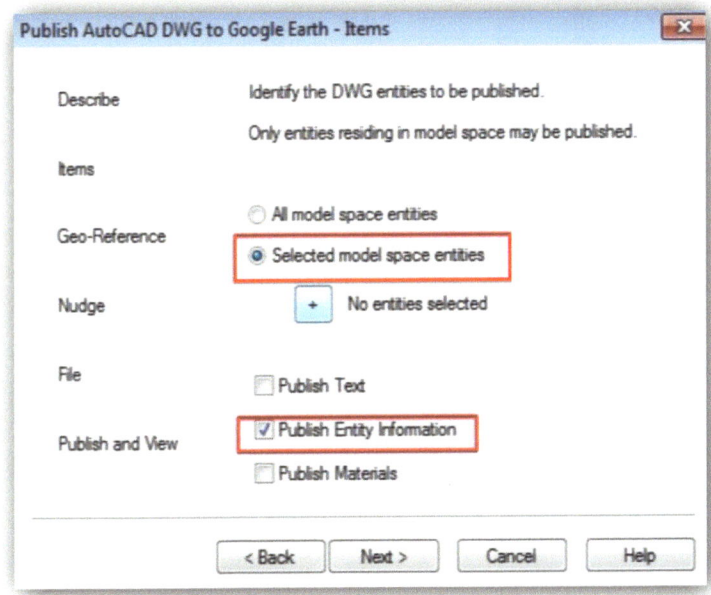

Select the parcel and Section line.

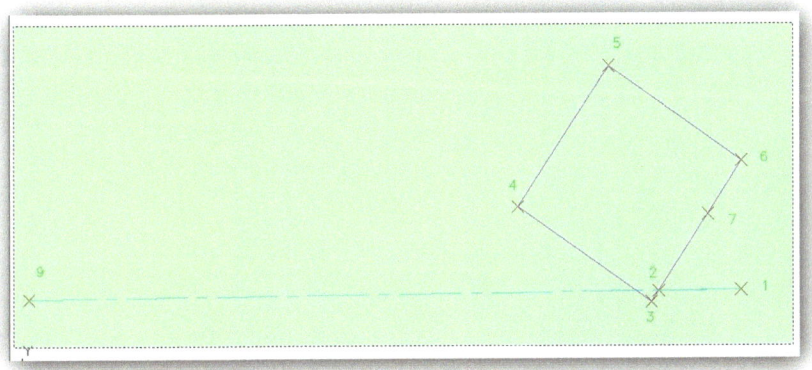

Notice the number of "entities" that have been selected. Click Next.

KML & KMZ

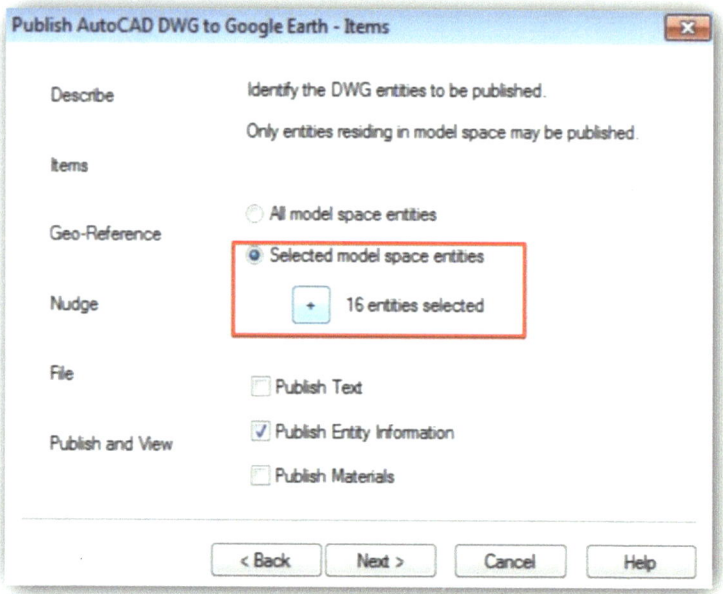

Select the "Drawing Coordinate System Transform". Notice the default value of your coordinate system is the same as what you had selected previously from the UNIT & ZONE tab under the "Drawing settings" menu.

KML & KMZ

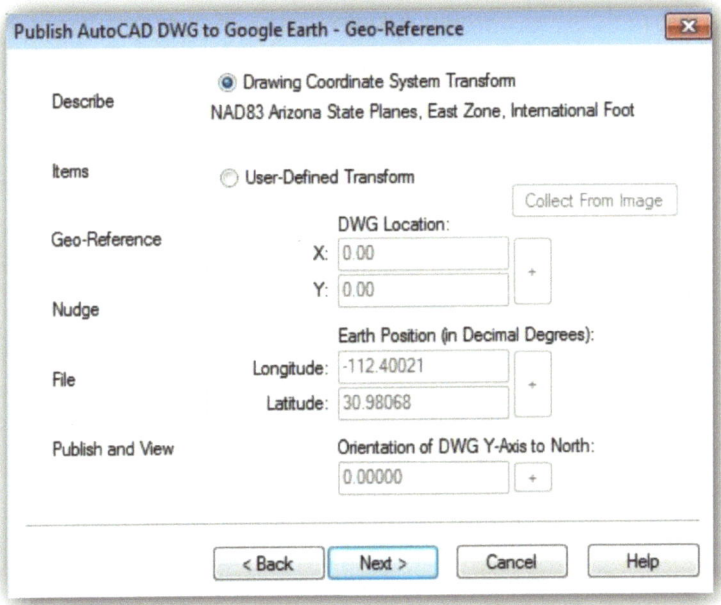

Now we'll decide how we would like the parcel and section line to be displayed. In most cases we will forego the "Nudge" option. Choose the "Drape Entities on ground" option. This will place the line work on the ground when shown on the GE aerial image regardless if any of the line work is elevated.

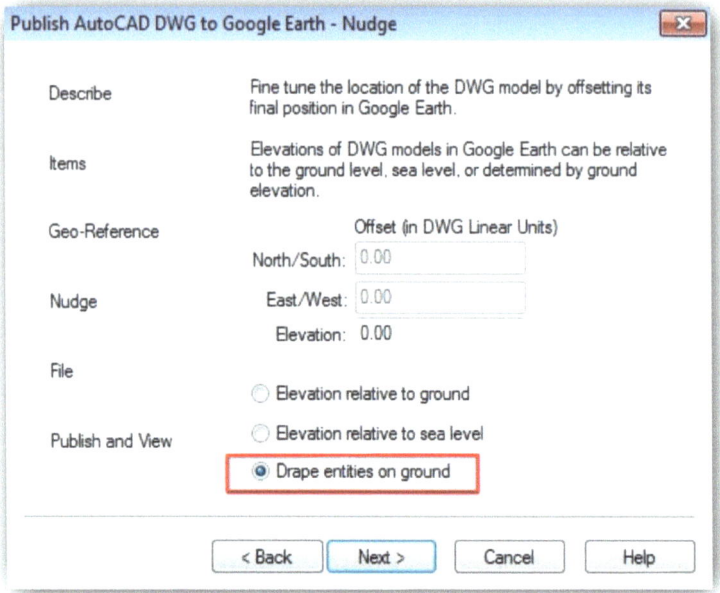

Save KMZ file

Now choose where to save the KMZ file. You can share this KMZ file with others as needed.

KML & KMZ

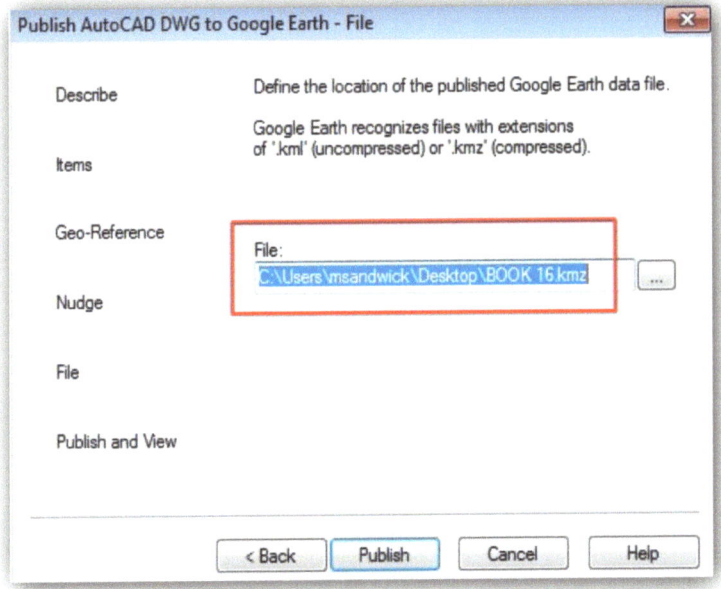

To immediately view the file in GE, click on the "Publish" button, then click "View" to send the KMZ file to GE which starts the Earth spinning and stopping at the parcel location.

If everything was entered correctly, GE will display the linework in the actual parcel location.

KML & KMZ

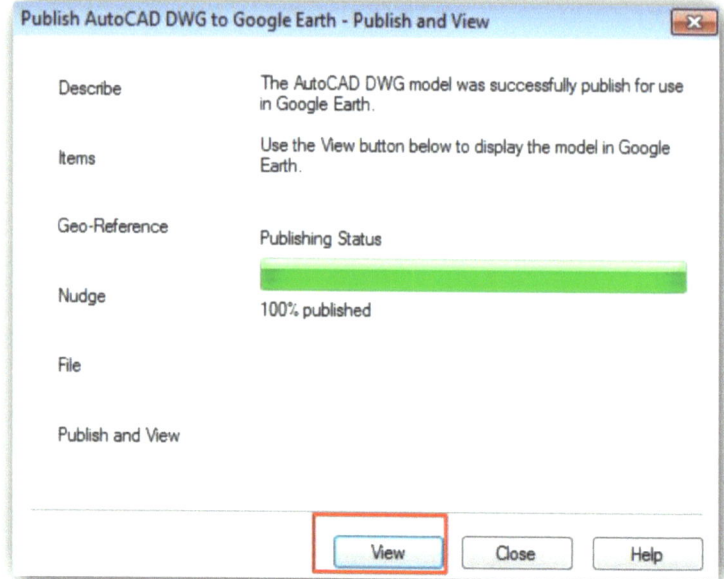

View in Google Earth

Enjoy your handiwork! One of the many values associated with publishing the project to GE, is to be able to visually check the spatial accuracy of your drawing. Does the line work show up where it is supposed to be? Keep in mind that not all line work will fit perfectly. In some instances, another GAF value can be utilized to commensurate GE models for displays and/or exhibits.

KML & KMZ

KML & KMZ

TRIMBLE BUSINESS CENTER

The following are the steps needed to create a KMZ file using TBC and open it in Google Earth.

- Convert Ground Coordinates to Grid Coordinates
- Create new TBC project
- Import Grid Coordinates
- Create a KMZ file
- Open the KMZ in Google Earth

Start by first converting the "Representative Parcel" Ground Coordinates to Grid Coordinates using the project GAF.

There are several programs available for converting Ground Coordinates to Grid Coordinates. If you have a large number of coordinate points, you will need to utilize a computer program to make this conversion.

The Representative Parcel only has a few coordinates to deal with so we will convert them using a simple spreadsheet.

Create a column for Ground (Northing and Easting) and enter the values for the Representative Parcel noted earlier in this book.

Next create a column for Grid (Northing and Easting). In these columns add the following formulas:

Grid Northing = Ground Northing / GAF

Grid Easting = Ground Easting / GAF

You should get the following:

KML & KMZ

Pt #	Northing - Gnd	Easting - Gnd	Northing - Grid	Easting - Grid
1	2157691.70913	1016642.48026	2157154.25415	1016389.24688
2 (POB)	2157688.31036	1016337.02636	2157150.85623	1016083.86906
3	2157649.39711	1016310.62333	2157111.95267	1016057.47261
4	2157985.95806	1015814.13046	2157448.42978	1015561.10341
5	2158482.66135	1016151.13398	2157945.00935	1015898.02299
6	2158147.63943	1016644.60747	2157610.07088	1016391.37356
7	2157959.09703	1016520.75779	2157421.57544	1016267.55473
8 (RP)	2162192.24047	1010281.87551	2161653.66446	1010030.22648
1A	2157662.35359	1014004.24314	2157124.90592	1013751.66691

Ground Coordinates to Grid Coordinates (GAF 1.00024915)

At this point you could save the Grid Coordinates into a comma delimited file (CSV) and import that into TBC. If you have a large number of coordinate points you would want to use a CSV file.

For this session, it is faster to add the handful of points directly into TBC. We will show you how that is done.

Start TBC then click on "Start a new project".

KML & KMZ

For the Representative Parcel, select the "Arizona East Zone, International Foot".

If you don't have the Arizona East Zone template in your list, then you will need to select the "International Foot" (read only) template then edit the project settings to change to the Arizona East Zone.

Consult the users manual on the steps required to edit the Project Settings.

KML & KMZ

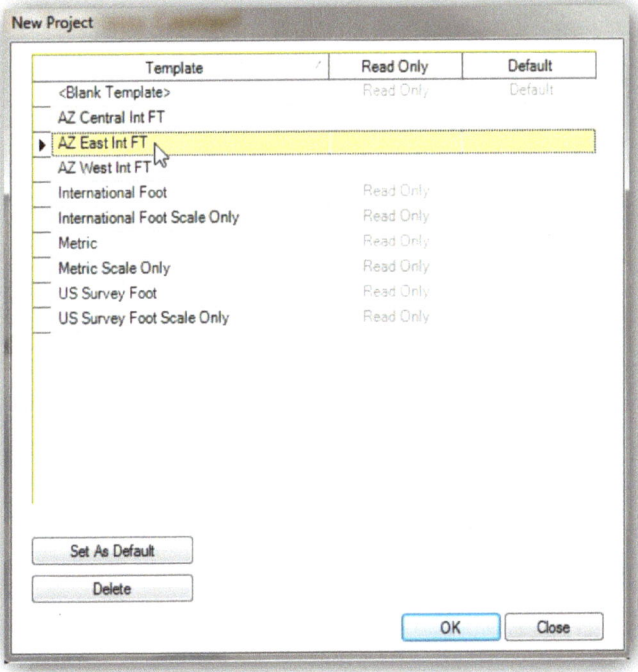

KML & KMZ

Select the "CAD" tab then click on to top half of "Create Point" icon.

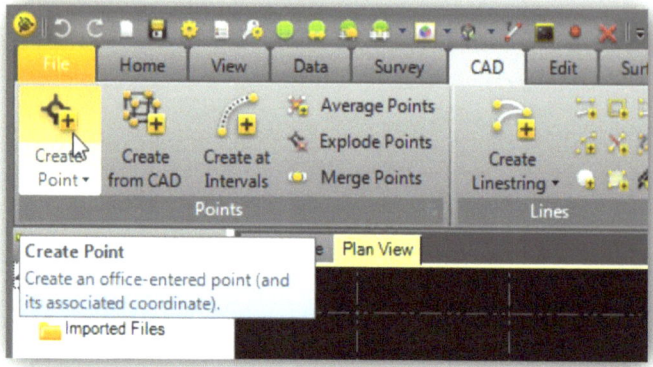

In the Create Point dialog box, type in "1" for the Point ID, then select "Grid" for the Coordinate Type.

Type in the Grid Northing for Point 1.

Type in the Grid Easting for Point 1.

Type in 5200 for the Elevation. An elevation must be provided in order to send the point to Google Earth. If you don't have an elevation for the point, use an average elevation for the project site.

Click the "Add" button to add this point to the Plan View.

Repeat this process for the remaining grid coordinates.

KML & KMZ

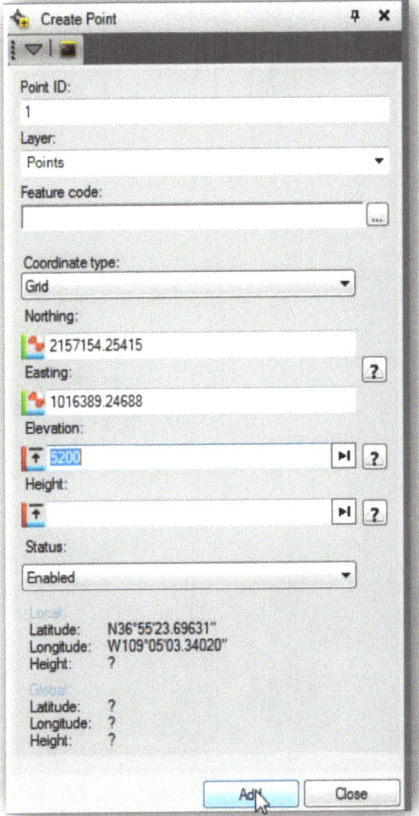

When all points have been added, click the "Close" button.

To display all of the points in the "Plan View" screen, select the "View" tab then click on the "Zoom Extents" icon.

KML & KMZ

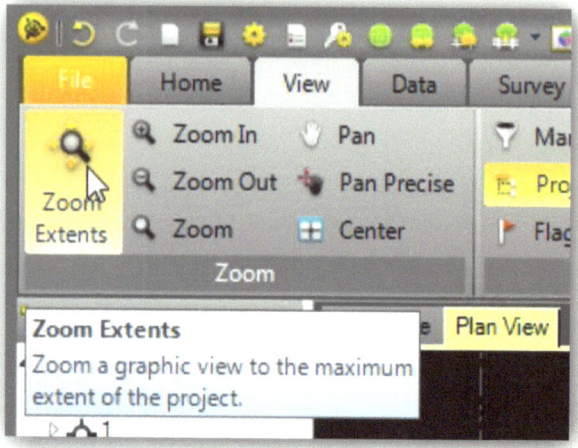

Your screen should look like the following:

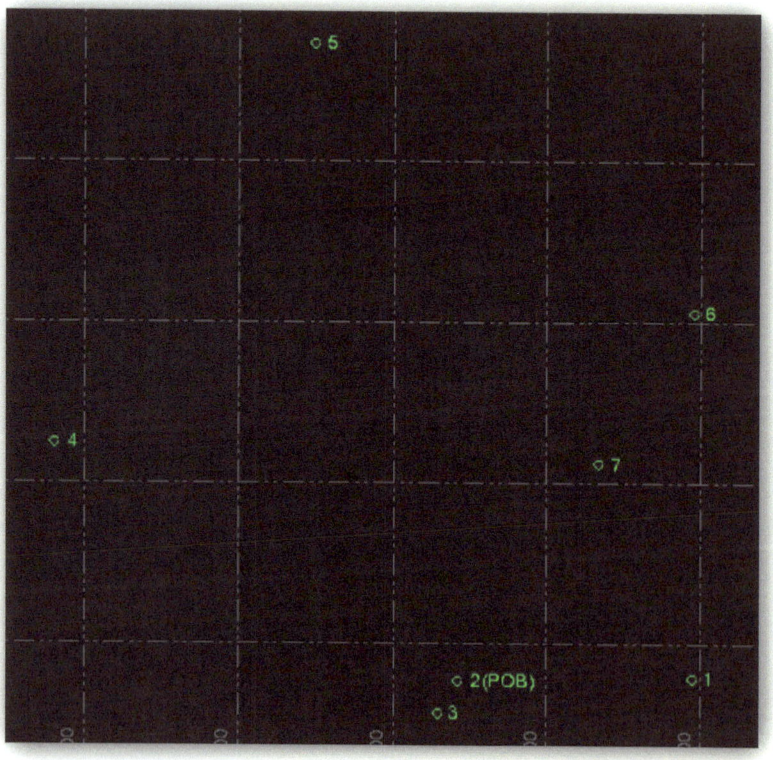

KML & KMZ

Select the "View" tab then click on the "Google Earth" icon under "Graphic Views".

Click on the "Options" button and click on "Select All".

Select "Clamp to ground" for the Altitude mode:

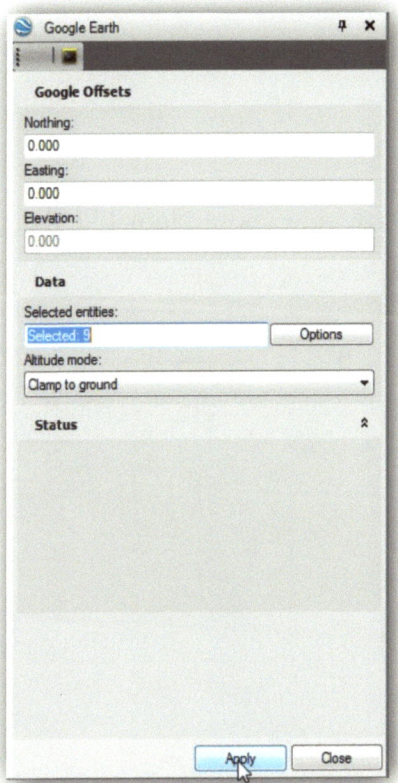

KML & KMZ

Click on the "Apply" button.

A KML file will be created and opened in Google Earth.

The Status box will display where the KML was saved. You will need to move this file to a location for safe keeping if you want to keep it.

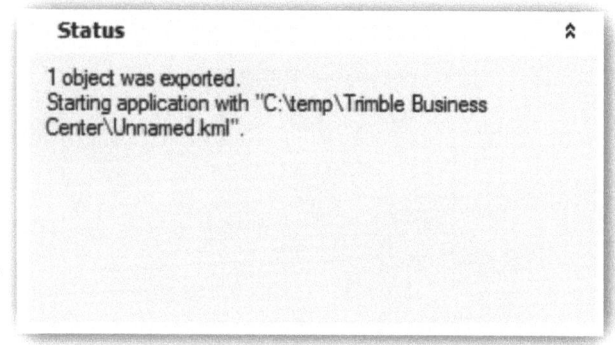

Google Earth view should look like the following:

KML & KMZ

You can click on any of the points to display the coordinates and elevation.

Within TBC you can add linework, set up layers, customize the point labels and so forth.

These step by step simple instructions are to quickly view points in Google Earth from TBC.

You can view a report for the latitude and longitude for each of the points in TBC.

Select the "File" tab then click on "Reports".

Select the "Point List" report. In the settings box on the right, select the features that you want in the report such as Local

KML & KMZ

coordinates, Combined scale factor, Meridian convergence angle, etc.

Finally click on the "Show Report" button at the bottom.

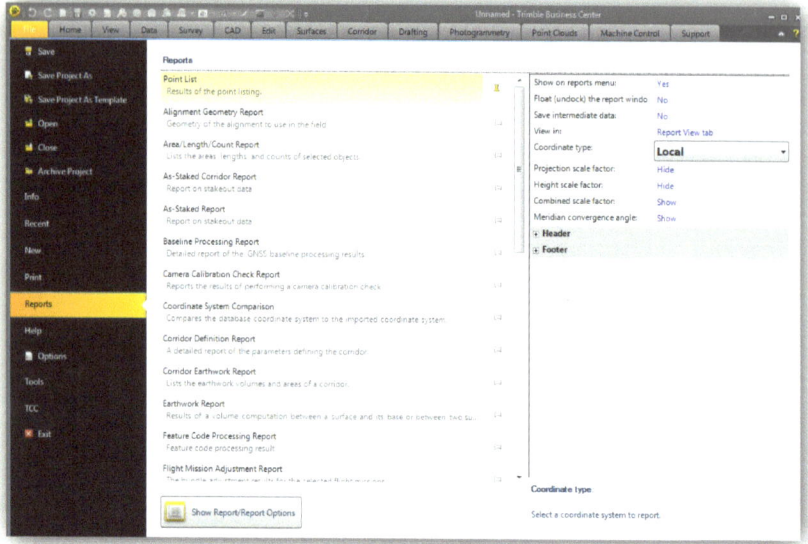

Here is a report that shows the selected information for each point.

KML & KMZ

Point List

ID	Latitude (Local)	Longitude (Local)	Height (Local) (International foot)	Feature Code	Combined Scale Factor	Meridian convergence angle
1	N36°55'23.69631"	W109°05'03.340 20"	5132.312		0.9997696283	0°39'01"
1A	N36°55'23.70090"	W109°05'35.823 58"	5132.338		0.9997677256	0°38'42"
2 (POB)	N36°55'23.69697"	W109°05'07.101 10"	5132.315		0.9997694072	0°38'59"
3	N36°55'23.31530"	W109°05'07.431 58"	5132.316		0.9997693880	0°38'59"
4	N36°55'26.69756"	W109°05'13.496 98"	5132.317		0.9997690293	0°38'55"
5	N36°55'31.56939"	W109°05'09.278 79"	5132.308		0.9997692731	0°38'58"
6	N36°55'28.20261"	W109°05'03.250 30"	5132.305		0.9997696302	0°39'01"
7	N36°55'26.35290"	W109°05'04.801 37"	5132.309		0.9997695403	0°39'00"
8 (RP)	N36°56'08.88753"	W109°06'21.029 33"	5132.310		0.9997650710	0°38'15"

9/14/2015 2:04:29 PM   Trimble Business Center

TBC is pretty robust with lots of bells and whistles.

This book only demonstrates a quick method of viewing points in Google Earth and not the many features associated with TBC.

For additional features within TBC, please consult the user manual.

ABOUT THE AUTHOR
Jim Crume P.L.S., M.S., CFedS

My land surveying career began several decades ago while attending Albuquerque Technical Vocational Institute in New Mexico and has traversed many states such as Alaska, Arizona, Utah and Wyoming. I am a Professional Land Surveyor in Arizona, Utah and Wyoming. I am an appointed United States Mineral Surveyor and a Bureau of Land Management (BLM) Certified Federal Surveyor. I have many years of computer programming experience related to surveying.

This ebook is dedicated to the many individuals that have helped shape my career. Especially my wife Cindy. She has been my biggest supporter. She has been my instrument person, accountant, advisor and my best friend. Without her, I would not be the professional I am today. Cindy, thank you very much.

Other titles by this author:

http://www.cc4w.net/ebooks.html

Follow us on Facebook

Books available on amazon.com

KML & KMZ

Printed - Digital - Apps
Many Titles to choose from.
www.cc4w.net

A **New** Math-Series of books with useful formulas, helpful hints and easy to follow step by step instructions.

Digital and **Printed Editions** Math-Series Training and Reference Books. Designed and written by Surveyors for Surveyors, Land Surveyors in Training, Engineers, Engineers in Training and aspiring Students.

www.facebook.com/surveyingmathematics

www.ingramcontent.com/pod-product-compliance
Lightning Source LLC
Chambersburg PA
CBHW040810200526
45159CB00022B/142